Microcomputer Math

William Barden, Jr., is a computer consultant specializing in small configuration computer systems. He has nearly 20 years experience in computer programming and computer systems analysis and design on a variety of computer systems. Mr. Barden is a member of the Association for Computing Machinery and the IEEE. His major interest is home computer systems. Amateur radio, mathematical games, and sailing are among his other interests.

Mr. Barden's books for SAMS include *How to Buy and Use Minicomputers and Microcomputers, How to Program Microcomputers, Microcomputers for Business Applications, Z-80 Microcomputer Handbook,* and *Z-80 Microcomputer Design Projects.*

Microcomputer Math

by

William Barden, Jr.

Howard W. Sams & Co., Inc.
4300 WEST 62ND ST. INDIANAPOLIS, INDIANA 46268 USA

International Standard Book Number: 0-672-21927-1
Library of Congress Catalog Card Number: 81-86554

Edited by: *Richard Krajewski*
Illustrated by: *T. R. Emrick*

Printed in the United States of America.

6-11-85

Preface

It isn't very long after the purchase of a microcomputer system that the microcomputer user encounters ominous references to "binary numbers," "hexadecimal value," "ANDing two numbers to get the result," or "shifting the result to multiply by two." Sometimes these references assume that the reader knows the binary system and operations; other times one gets the distinct feeling that the writer of the reference manual doesn't really know all that much about the operations either.

The purpose of *Microcomputer Math* is to remove some of the mystery surrounding the specialized math operations that are used in both BASIC language and assembly language. Such operations as binary number representation, octal and hexadecimal representation, two's complement operation, addition and subtraction of binary numbers, "flag" bits in microcomputers, logical operations and shifting, multiplication and division algorithms, multiple-precision operations, fractions and scaling, and floating-point operations are explained in detail, along with practical examples and exercises for self-testing.

If you can add, subtract, multiply, and divide decimal numbers, then you can easily perform the same operations in binary and other number bases, such as hexadecimal. This book will show you how.

This book makes an excellent companion to any course in assembly language or advanced BASIC.

Microcomputer Math is organized into ten chapters. Most of the chapters are based upon material contained in preceding chapters. Each chapter has self-test exercises at the end. It is to your benefit to go through the exercises, as they help fix the material in your mind, but we won't hold it against you if you use the book for reference only.

As you read, you'll notice some words are set in **boldface.** Most of these are computer terms that are defined in the Glossary. Using the Glossary, even the complete novice can easily learn and profit from this book.

The book is arranged as follows:

Chapter 1 discusses the binary system from the ground up and covers the conversions between binary and decimal numbers, while Chapter 2 describes octal and hexadecimal numbers, and conversions between these "bases" and decimal numbers. Hexadecimal numbers are used in both BASIC and assembly language.

Signed numbers and two's complement notation are covered in Chapter 3. Two's complement notation is used for negative numbers.

Chapter 4 discusses carries, overflow, and flags. These topics are used mostly in machine language or assembly language programming, but can be important in special BASIC programs also.

Logical operations, such as BASIC ANDS, ORS, and NOTS, are described in Chapter 5, along with the types of shifting possible in machine language. Then, Chapter 6 discusses multiplication and division algorithms, including both "unsigned" and "signed" operations.

Chapter 7 describes multiple-precision operations. Multiple precision may be used in both BASIC and assembly language to implement "unlimited precision" of any number of digits.

Chapter 8 covers binary fractions and scaling. This material is necessary to understand the internal format of floating-point numbers in BASIC.

Next, ASCII codes and ASCII conversions as they relate to numeric quantities are described in Chapter 9.

Finally, Chapter 10 provides an explanation of floating-point number representation as these numbers are used in many Microsoft BASIC interpreters.

Then, in the last section of the book, Appendix A contains the answers to the self-test questions, Appendix B is a glossary of terms, and Appendix C contains a listing of binary, octal, decimal, and hexadecimal numbers from 0 through 1023. The listing may be used for conversion from one type of number to the other. Finally, Appendix D contains a listing of two's complement numbers from −1 to −128, a handy reference that is not usually found in other texts.

WILLIAM BARDEN, JR.

To Dave Gunzel

A programmed instruction course that uses this book as a reference is available for many microcomputers from:

William Barden, Jr., Inc.
Post Office Box 3568
Mission Viejo, CA 92692

Send a self-addressed envelope for further information.

Contents

APPENDIX C

APPENDIX D

CHAPTER 1

The Binary System—
Where It All Begins

In the binary system, all numbers are represented by an "on/off" condition. Let's look at a quick example of binary in easy-to-understand terms.

BIG ED LEARNS BINARY

"Big Ed" Hackenbyte owns "Big Ed's," a restaurant that serves quick lunches and slow dinners in the middle of the area near San Jose, CA. This area, known as "Silicon Valley," has dozens of companies manufacturing microprocessor components for microcomputers.

Ed has eight serving people—Zelda, Olive, Trudy, Thelma, Fern, Fran, Selma, and Sidney. Depersonalization being what it is, they are assigned numbers for payroll reasons. The numbers assigned are:

	Number		Number
Zelda	0	Fern	4
Olive	1	Fran	5
Trudy	2	Selma	6
Thelma	3	Sidney	7

When Big Ed first implemented a "call board," he had eight lights, one for each of the serving people, as shown in Fig. 1–1. One day, though, Bob Borrow, a design engineer at a microprocessor components company known as Inlog, called Ed over.

"Ed, you could be a lot more efficient with your call board, you know. I can show you how we would have designed the board with one of our microprocessors."

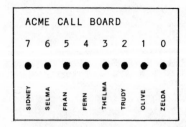

Fig. 1–1. Big Ed's call board.

Ed, being interested in the new technology, was receptive. The redesigned call board is shown in Fig. 1–2. It has three lights, controlled from the kitchen. When a serving person is called, a bell sounds. How is it possible to call any one of the eight serving people by lighting combinations of the three lights?

```
ACME  CALL  BOARD  MARK  II

    ●      ●      ●

●  BELL  TEST
```

Fig. 1–2. A binary call board.

"You see, Ed, this board is very efficient. It uses five less lights than your first call board. There are eight different combinations of lights. We really call them **permutations,** as there's a definite order to the light arrangement. I've prepared a chart of the permutations of the lights and the serving person called." He gave Ed the chart shown in Fig. 1–3.

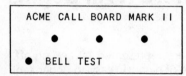

O	O	O	ZELDA	0
O	O	●	OLIVE	1
O	●	O	TRUDY	2
O	●	●	THELMA	3
●	O	O	FERN	4
●	O	●	FRAN	5
●	●	O	SELMA	6
●	●	●	SIDNEY	7

Fig. 1–3. Code chart for call board.

"There are only eight different permutations of lights, Ed—no more, no less. These lights are arranged in **binary** fashion. We use the binary system in our computers for two reasons. First of all, it saves parts. We reduced your number of lights from eight to three. Secondly, inexpensive computer components can usually represent only an on/off state, just as the lights are either on or off." He paused to sample some of his "Big Edburger." . . .

"I'll give these codes to my help to memorize," said Ed.

"Each serving person has only to memorize his unique code, Ed. I'll give you the key, so that you can decode which serving person is being called without reference to that chart."

"You see, each of the lights represents a power of two. The light on the right represents two to the zero power, the next light is two to the first power, and the leftmost light is two to the second power. This is really very similar to the decimal system, where each digit represents a power of ten." He scratched an example on the tablecloth, as shown in Fig. 1–4.

$$
\begin{array}{l}
3\ 7\ 5 \\
\quad 5 \times 10^0 = 5 \\
\quad 7 \times 10^1 = 70 \\
\quad 3 \times 10^2 = 300 \\
\hline
375
\end{array}
$$

Fig. 1–4. Comparison of decimal and binary notation.

$$
\begin{array}{l}
1\ 0\ 1 \\
\quad 1 \times 2^0 = 1 \\
\quad 0 \times 2^1 = 0 \\
\quad 1 \times 2^2 = 4 \\
\hline
5
\end{array}
$$

"Just as we can carry out the powers of ten to huge numbers, we can use as many powers of two as we want. We could use 32 lights, if we wanted. Now, to convert the three lights in binary to their decimal equivalent, add up the power of two for each light that is on." He scratched another figure on the tablecloth (Fig. 1–5).

"Well, that seems simple enough," Ed admitted. "From right to left, the lights represent 1, 2, and 4. If we had more waiters and waitresses, the lights would represent 1, 2, 4, 8, 16, 32, 64, 128 . . ." His voice trailed off, as he lost track of the next power of two.

"Exactly right, Ed. Typically, we have the equivalent of eight or sixteen lights in our microprocessors. We don't use lights, of course. We use semiconductor components that are either off or on." He scratched another figure on the tablecloth, which by now was overflowing with diagrams (Fig. 1–6).

"We call each of the eight or sixteen positions **bit positions.** 'Bit' is a contraction of the term 'binary digit.' After all, that's what we're talking about here, binary digits, just as decimal digits make up a decimal number. Your call board represents a 3-bit number."

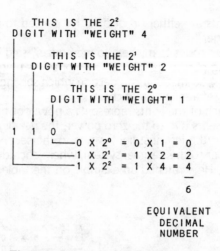

Fig. 1–5. Binary-to-decimal conversion.

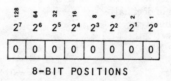

Fig. 1–6. An 8-bit and 16-bit representation.

"In eight bits we can represent any number between 0 and 1 + 2 + 4 + 8 + 16 + 32 + 64 + 128. Adding all of those up, we get 255, the largest number that can be held in eight bits."

"How about sixteen bits?" asked Ed.

"You figure it out," said Bob. "I've got to get back to the job of designing microprocessors."

Ed drew up a list of all the powers of two up to fifteen. He then added them together to come up with the result shown in Fig. 1–7, a total of 65,535—the largest number that can be held in sixteen bits.

"This microprocessor business is easy," Ed said with a grin, as he bit into his Big Ed's Jumboburger with an 8-bit byte.

WEIGHT	BIT POSITION
1	0
2	1
4	2
8	3
16	4
32	5
64	6
128	7
256	8
512	9
1024	10
2048	11
4096	12
8192	13
16384	14
32768	15
65535	

Fig. 1–7. A 16-bit maximum value.

MORE ON BITS, BYTES, AND BINARY

Bob Borrow's explanation of binary pretty well capsulizes microcomputer binary representation. The basic unit is a bit, or binary digit. A bit can be either on or off. Because it's much easier to write a "0" or "1," these digits are used in place of "on" or "off" when we're representing binary values.

The **bit position** of the binary number refers to the position of the bit in the binary value. The bit position in most microcomputers has a number associated with it, as shown in Fig. 1–8. As bit positions really represent a power of two, it's convenient to number them according to the power of two represented. The rightmost bit is two to the zero power, and the bit position is therefore zero. The bit positions moving to the left are numbered 1 (two to the first power), 2 (two to the second power), 3, 4, 5, and so forth.

In all current microcomputers, a collection of eight bits is called a **byte.** It's somewhat obvious to see how this evolved if one imagines early computer engineers having lunch at the equivalent of Big Ed's.

Microcomputer memory is often specified in the number of bytes

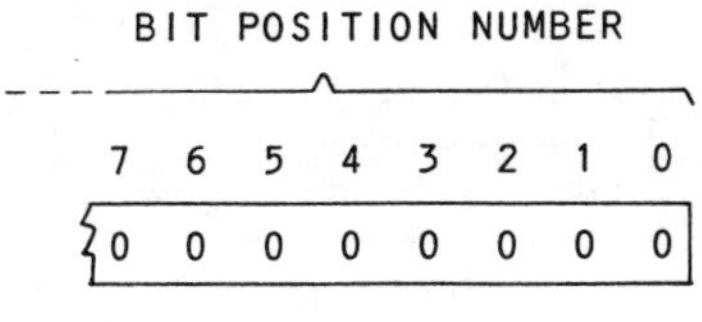

BIT POSITION NUMBER

7	6	5	4	3	2	1	0
0	0	0	0	0	0	0	0

Fig. 1–8. Bit-position numbering.

that it represents. Each byte roughly corresponds to one character, as we shall see in later chapters. Operations to and from memory are generally made a byte at a time.

Registers in the microprocessor portion of a microcomputer are also one or two bytes in length. Registers are really nothing more than fast-access memory locations that are used for temporary storage in the microprocessor. There are typically ten bytes of register storage in the microprocessor and up to 65,535 bytes in the memory of a microcomputer, as illustrated in Figs. 1–9 and 1–10.

Are there any other groupings above or below bytes? Some microcomputers talk about **words,** which may be two or more bytes but, generally, bytes are the most commonly used collection of bits we talk about. Sometimes, groupings of four **bits** are referred to as a **nibble.** (Early computer engineers must have been obsessed with food!)

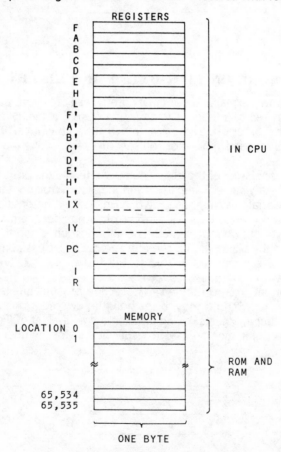

Fig. 1–9. Z-80 registers and memory.

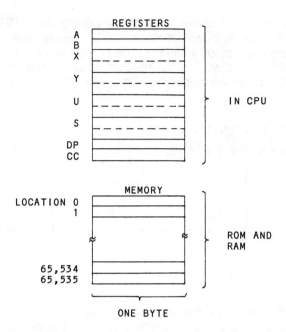

Fig. 1–10. 6809E registers and memory.

BASIC generally works with eight or sixteen bits of data at a time—one or two bytes. One byte is used for the BASIC PEEK or POKE instructions, which allow the user to read or write data into a memory location. This is valuable when used to change some of the system parameters that otherwise would be inaccessible from BASIC.

Two bytes are used for **integer variable** representation. In this type of format, a BASIC variable can hold values of −32,768 through +32,767. This number is held as a **signed** integer value, which we'll look at in Chapter 3.

If you're interested in assembly language on your microcomputer, then you will be working with single bytes to perform arithmetic operations such as addition and subtraction, and with one to four bytes to represent the **machine language** equivalent of the instruction to the microprocessor.

Because we'll be continually referencing one- and two-byte values, let's investigate them further.

CONVERTING FROM BINARY TO DECIMAL

Big Ed found the maximum values that could be held in one and two bytes by adding up all of the powers of two. They were 255 for a one-byte value, and 65,535 for a two-byte value. Any number in between can be represented by setting the appropriate bits.

Suppose that we had the 16-bit binary value 0000101001011101. To find the equivalent decimal number of this binary value, we could use Big Ed's method of adding up the powers of two. This is done in Fig. 1–11, where we find the result of 2653. Is there an easy way to convert from binary to decimal? Yes, there are a number of easier ways.

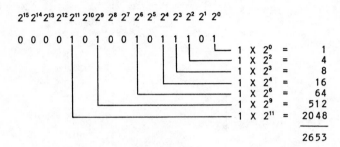

Fig. 1–11. Binary-to-decimal conversion.

The first way is by reference to a table of values. We've included such a table in Appendix C, which shows the relationship between binary, octal, decimal, and hexadecimal for values up to 1023. As displaying values up to 65,535 for sixteen bits takes up quite a bit of space, there must be a more efficient way.

A method that works surprisingly well is called **"double-dabble."** After a little bit of practice, it becomes very easy to convert ten or so bits, in binary, to decimal. In the next chapter, we'll show you a method that works for sixteen bits or more. Double-dabble is a strictly mechanical procedure that automates the conversion process. It's harder to describe than do. To use it, do the following:

1. Find the first (leftmost) 1 bit of the binary number.
2. Multiply the 1 bit by two and add the next (rightmost) bit of 0 or 1.
3. Multiply the result by two and add this result to the next bit of 0 or 1.
4. Repeat this process until the last (rightmost) bit has been added to the total. The result is the decimal equivalent of the binary number.

This procedure is shown in Fig. 1–12 for the quantity of 2653 used in the previous example.

When you've been working with binary numbers for some time, you'll probably immediately recognize 1111 as decimal 15 and 11111 as 31. There are also little tricks that you may employ, such as realizing

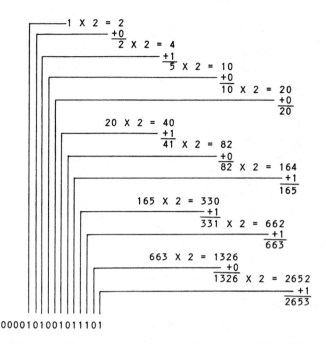

Fig. 1–12. Binary-to-decimal conversion using double-dabble.

that 11110 is one less than 31, or that 101000 is actually 1010 = 10 times 4 for a value of 40. For the time being, however, just do a few examples to get used to the procedure; proficiency will automatically come with time, and it probably isn't worth it to spend dozens of hours becoming a binary whiz.

CONVERTING FROM DECIMAL TO BINARY

How about conversion in the opposite direction? This is a somewhat different process. Suppose that we have the decimal number 250 to convert to binary. We'll look at several methods.

The first method is the "inspection/powers of two" method. It could also be called successive subtraction of powers of two, but Big Ed might not like the name. In this method, all we do is to try to subtract a power of two and put a 1 bit into the proper bit position, if we can (see Fig. 1–13). Some of the powers of two are: 256, 128, 64, 32, 16, 8, 4, 2, and 1, starting with the larger powers. It's obvious that 256 won't go, so we'll put a 0 in that bit position. The power 128 does go, leaving 122. We put a 1 into bit-position 7. The power 64 goes into 122, leaving 58, so we put a 1 into bit-position 6. This process is repeated until the last bit position has been calculated, as shown in Fig. 1–13.

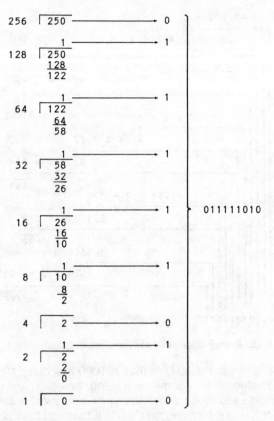

Fig. 1-13. Decimal-to-binary conversion by inspection.

The above method is pretty tedious. Is there a better way? One better method is called the "divide by two and same remainders" method. In this method, we do exactly what the name implies; it is illustrated in Fig. 1–14. The first division is 2 into 250, yielding 125, remainder 0. The next division is 2 into 125, yielding 62, remainder 1. The process is repeated until the "residue" is 0. Now the remainders are arranged in reverse order. The result is the binary equivalent of the decimal number.

PADDING OUT TO EIGHT OR SIXTEEN BITS

Here's a point that may not be apparent. Unused bit positions to the left are filled with 0 bits. This is simply a nicety used to "pad out" the binary number to eight or sixteen bits, whichever size we're working with at the time. It has some implications for **signed numbers,** however, so it's best to start using the practice early in the game.

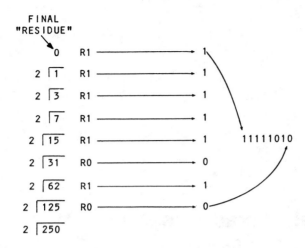

Fig. 1–14. Decimal-to-binary conversion by "divide and save remainders" method.

In the next chapter, we'll cover the shorthand notation of **octal** and **hexadecimal** numbers. In the meantime, try your hand at some self-test exercises for practice in conversion between decimal and binary numbers.

EXERCISES

1. List the binary equivalents of decimal 20 through 32.

2. Convert the following binary numbers to their decimal equivalents: 00110101, 00010000, 01010101, 11110000, 0011011101101001.

3. Convert the following decimal numbers to their binary form: 15, 26, 52, 105, 255, 60000.

4. "Pad out" the following binary numbers to eight bits: 101, 110101, 010101.

5. What is the largest decimal number that can be held in four bits? Six bits? Eight bits? Sixteen bits? If n is the number of bits, what general statement can you make about the largest number that can be held in n bits? (The response, "Some mighty big numbers!", is not considered acceptable.)

CHAPTER 2

Octal, Hexadecimal, and Other Number Bases

Hexadecimal and octal are variants of binary numbers. They are commonly used in microcomputer systems, especially hexadecimal. Data can be specified in hexadecimal notation in both BASIC and assembly language on many microcomputers. We'll discuss hexadecimal and octal in this chapter, along with some other interesting number systems. Let's revisit Big Ed . . .

CHILI IS BIG ON REGULUS

"We don't get many of your type around here," said Big Ed, as he set down a bowl of Big Ed's Chili Surprise in front of a male customer with scaly green skin.

"I know I should say something like 'Yes, and at these prices, you'll get even fewer,' but I'll tell you the truth. I'm just in from the United Nations to look over your semiconductor industry," said the visitor. "It's very impressive."

"I've done some work in it myself," said Big Ed, glancing at his call board. "Where you from, Buddy? I couldn't help noticing your eight-fingered hands."

"You've probably never heard of it—it's just a small star . . . , er . . . , place. These hands, by the way, are what bring me to Silicon Valley. Your recent microprocessor products are a natural for my people. Would you like to hear about it?"

Ed nodded.

"You see, my people are called Hackers. We base everything upon powers of sixteen." (He glanced at Big Ed to see if he had gotten the

24

pun.) "When our civilization was first developed, we counted on our hands. We found it pretty easy to count up to your value of sixteen by simply using our fingers. Later, we needed to express larger numbers. Eons ago, one of our kind discovered **positional notation.** As we have sixteen fingers, our numbers use a base of sixteen, just as your numbers use a base of ten. Each digit position represents a power of sixteen—1, 16, 256, 4096, and so forth."

"Exactly how does that work?" said Big Ed, anxiously eyeing the fresh tablecloth he had put on several days ago.

"Here's a typical number," said the visitor, swiftly scratching on the tablecloth. "Our number A5B1 represents . . ."

$$A \times 16^3 + 5 \times 16^2 + B \times 16^1 + 1 \times 16^0$$

"But, what are the As and Bs?" asked Big Ed.

"Oh, I forgot. When we count on our fingers, we count 1, 2, 3, 4, 5, 6, 7, 8, 9, A, B, C, D, E, and F. Actually, that's only fifteen fingers—our last represents the number after 'F,' our '10,' which is your sixteen."

"Aha! So A represents our decimal 10, B represents decimal 11, and so forth," said Ed, drawing a chart as shown in Fig. 2–1.

DECIMAL	BASE 16
0	0
1	1
2	2
3	3
4	4
5	5
6	6
7	7
8	8
9	9
10	A
11	B
12	C
13	D
14	E
15	F

Fig. 2–1. Base-16 representation.

"Exactly right!" said the visitor. "Now you can see that our number A5B1, in our base 16, is equal to your number 42,417."

$$A \times 16^3 + 5 \times 16^2 + B \times 16^1 + 1 \times 16^0 =$$
$$10 \times 4096 + 5 \times 256 + 11 \times 16 + 1 \times 1 =$$
$$42,417$$

"Well, to make a long story short, your decimal number system is really terrible. We almost decided not to trade with you after we discovered that you used decimal numbers. Fortunately, however, we found that the number base you most frequently used on your computers was hexadecimal, which is our base 16."

"Wait a minute!" cried Big Ed. "We use **binary** on our computer systems!"

"Well, yes, of course, on a 'hardware' level. But you use hexadecimal as a kind of shorthand to represent data values and instructions in memory. After all, binary and hexadecimal are almost identical." He went on after seeing Big Ed's puzzlement.

"Look, suppose that I have the number A1F5. I'll represent it by holding up my hands (Fig. 2–2). The eight fingers of the hand on your right represent the two hexadecimal digits of F5. The eight fingers on your left represent the two hexadecimal digits of A1. Each hexadecimal digit is represented by four fingers, which hold four bits, or the digit in binary. Get it?"

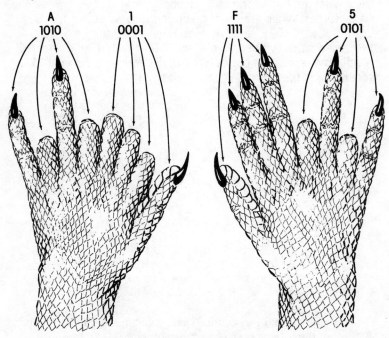

Fig. 2–2. Hexadecimal shorthand.

Big Ed scratched his head. "Let's see, 5 in binary is 0101, and that's represented by those four fingers. The next group of four represents the F, which is really 15, or binary 1111. The next group . . . , Oh, I get it. Instead of writing down 1010000111110101, you just write a shorthand notation of A1F5."

"Big Ed, you don't just serve the best chili this side of Altair, you're a mathematician to boot!" exclaimed the visitor, as he left with a scaly green eight-fingered wave.

Big Ed looked at the sixteen-sided copper piece that had been left as a tip and then started figuring on the tablecloth. . . .

HEXADECIMAL

Ed's visitor was correct. Hexadecimal is used on microcomputers because it is a convenient way to shorten long strings of binary ones and zeroes. Each group of four bits can be converted to a hexadecimal (we'll use **hex** from now on) value of 0, 1, 2, 3, 4, 5, 6, 7, 8, 9, A, B, C, D, E, or F, as shown in Table 2–1.

Table 2–1. Binary, Decimal, and Hexadecimal Representation

Binary	Decimal	Hex
0000	0	0
0001	1	1
0010	2	2
0011	3	3
0100	4	4
0101	5	5
0110	6	6
0111	7	7
1000	8	8
1001	9	9
1010	10	A
1011	11	B
1100	12	C
1101	13	D
1110	14	E
1111	15	F

It's a simple matter to convert from binary to hex. Starting from the rightmost bit (bit 0), divide the binary number into groups of four. If you do not have an integer multiple of four (4, 8, 12, etc.), you may have some bits left over on the left; in this case, simply pad out with zeroes. Now, convert each group of four bits into a hex digit. The result is the hexadecimal representation of the binary value, which is one-fourth as long in terms of characters. An example is shown in Fig. 2–3.

To convert from hex to binary, do the inverse. Take each hex digit and convert it into a 4-bit group. An example is shown in Fig. 2–4.

Hex notation is used for the Z-80 microprocessor, the 6502 microprocessor, the 6809E microprocessor, and many others.

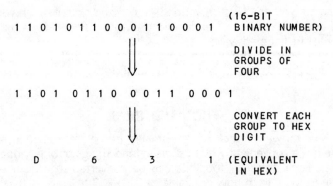

Fig. 2–3. Binary to hex conversion.

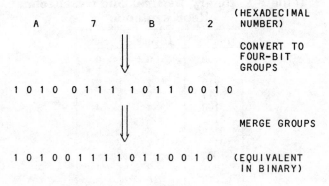

Fig. 2–4. Hex to binary conversion.

Converting Between Hex and Decimal

Converting between binary and hex is easy. What about decimal and hex conversions? Many of the principles and techniques discussed in the first chapter still apply.

To convert from hex to decimal by the powers of sixteen method, take each hex digit and multiply by the appropriate power of sixteen, as Big Ed did. To convert 1F1E, for example, we'd have:

$$1 \times 4096 + 15 \times 256 + 1 \times 16 + 14 \times 1 = 7966$$

The double-dabble method can also be adapted to a hexa-dabble. (In fact, this scheme works for any number base.) Take the leftmost hex digit and multiply by 16. Add to the next hex digit. Multiply the result

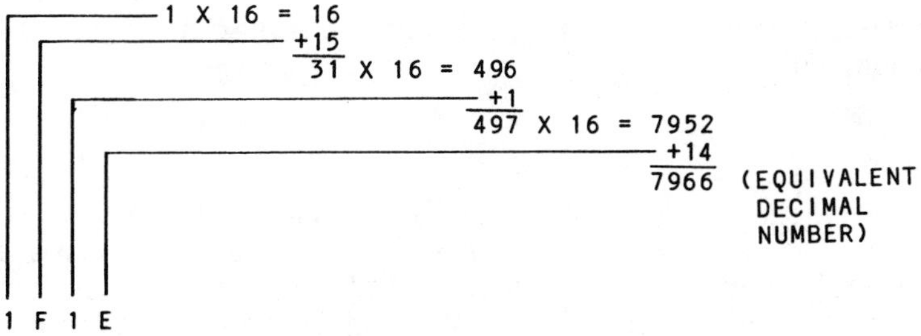

Fig. 2–5. Hexa-dabble hex-to-decimal conversion.

by 16. Add the next hex digit. Repeat the multiplication and addition process until the last (rightmost) hex digit has been used. This procedure is shown in Fig. 2–5.

You might try this procedure and compare the results with a double-dabble of the binary number. (Needless to say, the results had better be the same.)

Converting From Decimal to Hexadecimal

The method of successive subtraction of powers of sixteen doesn't work too well here, as you might have to do fifteen subtractions to get one hex digit! The analogous "divide by sixteen, save remainders" method, however, works quite well as shown in Fig. 2–6. Take the decimal value 48,555 as an example. Dividing by 16 yields a value of 3034 with a remainder of 11 (hex B). Then, dividing the 3034 by 16 yields 189 with a remainder of 10 (hex A). Dividing the 189 by 16 yields 11, with a remainder of 13 (hex D). Finally, dividing 11 by 16 gives 0,

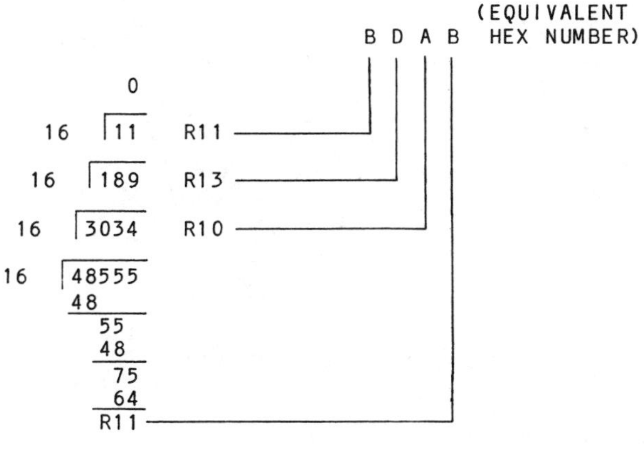

Fig. 2–6. Decimal-to-hex conversion.

with a remainder of 11. The remainders in reverse order are the hex equivalent, BDAB.

OCTAL

Hexadecimal is the most commonly used number base on microcomputers. However, octal, or base 8, is also sometimes used. Octal is used primarily on the 8080 microprocessor. As the Z-80 microprocessor used in many microcomputers is an offspring of the 8080, many of the instructions of the 8080 are used by the Z-80. Some of these instructions use fields that are made up of three bits and positioned so that octal representation is convenient.

Octal values use powers of 8; that is, position 1 is to the zero power (8^0), position 2 is to the first power (8^1), position 3 is to the second power (8^2), and so forth. Here, there is no problem with assigning names to new digits, as there was in the case of the hex digits A through F. The octal digits are 0, 1, 2, 3, 4, 5, 6, and 7. Each digit can be represented by three bits.

Converting Between Binary and Octal

Conversion between binary and octal numbers is similar to hex conversion. To convert from binary to octal, group the bits into 3-bit groups, starting at the right. If you are working with 8- or 16-bit numbers, you will have some bits left over. Pad these with zeroes. Now change each group of three bits into an octal digit. An example is given in Fig. 2–7. To convert from octal to binary, reverse the process.

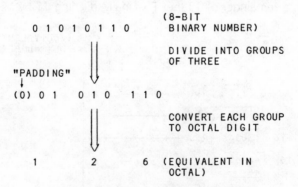

Fig. 2–7. Binary-to-octal conversion.

Converting Between Octal and Decimal

Conversion from octal to decimal may be handled by the powers of eight method or by octal-dabble. In the first method, multiply the octal digit by the power of eight. To convert octal 360, for example, we'd have:

$$3 \times 8^2 \quad + 6 \times 8^1 \quad + 0 \times 8^0 =$$
$$3 \times 64 \quad + 6 \times 8 \quad + 0 \times 1 \quad =$$
$$192 \quad + 48 \quad + 0 \quad = 240$$

In using the octal-dabble method, go to the leftmost octal digit, and multiply it by eight. Add the result to the next octal digit. Multiply this result by eight, and add to the next octal digit. Repeat this process until the last (rightmost digit) has been added. This procedure is shown in Fig. 2–8.

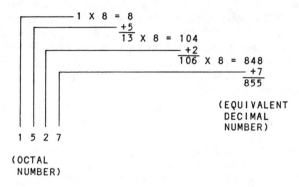

Fig. 2–8. Conversion from octal to decimal using octal-dabble.

To convert from decimal to octal, the "divide and save remainders" technique is used again. Divide the octal number by eight and save the remainder. Divide the result again, and repeat until the remainder is

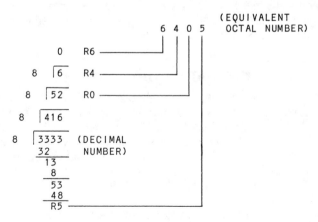

Fig. 2–9. Decimal-to-octal conversion by the "divide and save remainders" method.

less than eight. The remainders in reverse order are the equivalent octal number. An example is shown in Fig. 2–9.

WORKING IN OTHER NUMBER BASES

Although hexadecimal and octal are the most popular number bases used in microcomputers, it's possible to work in any number base whether it is base 3, base 5, or base 126. A popular software package from Microsoft Consumer Products, muMath®, allows just about any base to be used for a large number of digits of **precision.**

Two examples of the use of such a base might be interesting. An old technique of compressing three characters into two bytes uses **base 40.** Each of the alphabetic characters A through Z, the digits 0 through 9, and four special characters (period, comma, question mark, and exclamation mark, or any four others) are assigned a code of 0 through 39 decimal. As the largest three-digit base-40 number is $39 \times 40^2 + 39 \times 40^1 + 39 \times 40^0$, or 63,999, the three base-40 digits can be nicely put into sixteen bits (two bytes). Normally, the three characters would have to be held in three bytes. This "compression" results in a 50% saving in memory space for text using only A through Z, 0 through 9, and four special characters.

A second example would be the special processing of a tic-tac-toe array. Each of the nine elements of a tic-tac-toe "square" holds a space, a "naught," or a "cross." As there are three characters, a base 3 representation (space = 0, naught = 1, cross = 2) may be used to advantage. The largest base 3 number for this encoding would be $2 \times 3^8 + 2 \times 3^7 + 2 \times 3^6 + 2 \times 3^5 + 2 \times 3^4 + 2 \times 3^3 + 2 \times 3^2 + 2 \times 3^1 + 2 \times 3^0$, or 19,682, which again can be held nicely in sixteen bits or two bytes.

Numbers in other bases may be converted to decimal numbers and back again by the "base-dabble" method and the "divide by base, save remainders" method, in a similar fashion to octal and hexadecimal numbers.

STANDARD CONVENTIONS

In the remainder of this book, we'll occasionally use the suffix "H" for hexadecimal numbers. The number "1234H," for example, will mean 1234 hexadecimal and not 1234 in decimal. (It will also be equivalent to &H1234 in certain versions of BASIC.) Also, we'll express powers in the same way BASIC does. Instead of "superscripting," we'll use an up arrow (↑) to represent exponentiation—raising a number to a power. The number 2^8 will be represented by 2 ↑ 8, 10^5 by 10 ↑ 5, and 10^{-7} (or $\frac{1}{10}^7$) by 10 ↑ − 7.

In the next chapter we'll consider **signed numbers.** In the meantime, work through the following exercises in hexadecimal, octal, and special bases.

EXERCISES

1. What does the hexadecimal number 9E2 represent in terms of powers of 16?

2. List the hexadecimal equivalents of decimal 0 through 20.

3. Convert the following binary numbers to hexadecimal: 0101, 1010, 10101010, 01001111, 1011011000111010.

4. Convert the following hexadecimal numbers to binary: AE3, 999, F232.

5. Convert the following hexadecimal numbers to decimal: E3, 52, AAAA.

6. Convert the following decimal numbers to hexadecimal: 13, 15, 28, 1000.

7. The greatest memory address in a 64K microcomputer is 65,535. What is this in hexadecimal?

8. Convert the following octal numbers to decimal: 111, 333.

9. Convert the following decimal numbers to octal: 7, 113, 200.

10. What can you say about the octal number 18? (Limit your answers to 1000 words or less, please.)

11. In a number system based on seven, what would be the decimal equivalent of 636, base 7?

CHAPTER 3

Signed Numbers and Two's Complement Notation

Thus far we've talked about **unsigned binary numbers** that represented only positive values. In this chapter, we'll learn how to represent positive **and** negative values in microcomputers.

BIG ED AND THE BINACUS

"Hi guy. What'll it be?" Big Ed asked a slight customer with a long queue and colorful brocaded coat.

"I'll have the chop suey," said the customer.

"What do you have there?" Big Ed asked, spying a strange looking device that resembled an abacus with only a few beads. "It looks like an abacus."

"No, it's a **binacus**!" said the customer. "It was to be my road to fame and fortune, but alas, he who looks for those two companions on the road of life will find only chuckholes. Would you care to hear about it?"

Ed threw up his hands, knowing he had no choice; it was a slow afternoon. . . .

"This device is like an abacus, but it deals with binary numbers. That's why it is sixteen columns wide and has only one bead per column (Fig. 3–1). I spent years developing it, and only recently tried to market it through some of the microcomputer firms here. They all rejected it!"

"Why, pal?" said Ed, handing over the chop suey.

"It can represent any number from 0 through 65,535 and one can easily add or subtract binary numbers with it. But it has no way of representing negative numbers!"

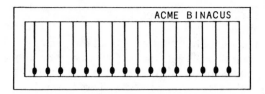

Fig. 3–1. The Binacus.

Just as he uttered those words, Bob Borrow, the Inlog computer engineer, walked in. "I couldn't help hearing your tale," he said. "I think I might be of some help."

"You see, your problem is one that computer engineers faced many years ago. Many times, it's sufficient to simply hold an absolute number. For example, most microprocessors can address 65,536 separate addresses, numbered from 0 through 65,535. There's no reason to have negative numbers in this case. On the other hand, microcomputers have the ability to perform arithmetic operations on **data** and must have the ability to hold both negative and positive values."

"There are different schemes for representing negative numbers. Look at this." Ed winced as Bob reached for the tablecloth.

"You could simply add an extra bead, a seventeenth, on the left end of the binacus. If the bead was up, the number represented would be a positive number; if the bead was down, the number would be a negative number. This scheme is called **sign/magnitude** representation." (See Fig. 3–2.)

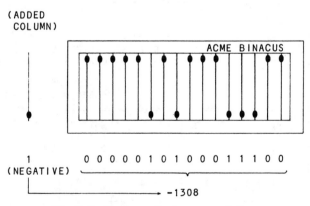

Fig. 3–2. Sign/magnitude representation.

"That's true," said the stranger. "Of course, it would mean some redesign," he mused, looking at the invention.

"Wait, I haven't come to my point. I've got a scheme for you that won't require any redesign at all! It's called two's complement notation, and it allows you to hold numbers from $-32,768$ to $+32,767$ without any modification."

"Oh, sir, if you could do that . . . ," said the stranger.

"In this scheme, we'll let the sixteenth bead on the end, in bead position 15, represent the **sign bit.** If the bead is up or 0, the sign will be positive, and the remaining beads will hold the value of the number. Since we'll have only 15 beads, the number represented will be from 0 (000 0000 0000 0000) through 32,767 (111   1111   1111 1111)."

"What about negative numbers?" asked the stranger.

"I was getting to that. If the bead in bead position 15 is down, or 1, then the number represented is a negative number. In this case, flip all of the beads that are up, or 0, down to a 1. Flip all of the beads that are down, or 1, up to a 0. Finally, add one."

"Thanks for your trouble, sir," said the stranger, with startled eyes, as he picked up the device and started to walk out the door.

"No, wait, this system works!" cried Bob. "Here, let me show you. Suppose that you have the configuration 0101 1110 1111 0001.

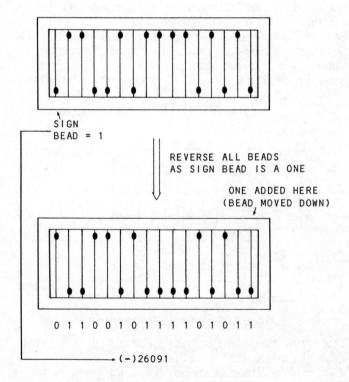

Fig. 3–3. Two's complement notation.

The sign bead is a 0, so the number represents 101 1110 1111 0001 (hexadecimal 5EF1 or decimal 24,305). Now, suppose that the configuration is 1001 1010 0001 0101. The sign bead is a one, so the number is negative. Now, reverse all the beads and add one." (See Fig. 3–3.)

"The result is 0110 0101 1110 1011. We now convert the number to decimal 26,091; therefore, the number represented is a −26,091. This scheme is the same one that microcomputers use. You'll easily be able to sell your binacus idea to a manufacturer if you tell them it uses this scheme of **two's complement** notation for negative numbers!"

The stranger looked dubious. "Let's see if I understand this. If the sign bead is a one, I reverse all the beads and add one? Let me try it with a few examples." He moved the binacus beads and calculated on the tablecloth. Ed grimaced. A chart of what he came up with is shown in Fig. 3–4.

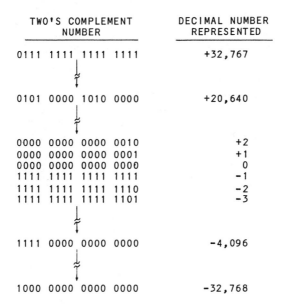

TWO'S COMPLEMENT NUMBER	DECIMAL NUMBER REPRESENTED
0111 1111 1111 1111	+32,767
0101 0000 1010 0000	+20,640
0000 0000 0000 0010	+2
0000 0000 0000 0001	+1
0000 0000 0000 0000	0
1111 1111 1111 1111	−1
1111 1111 1111 1110	−2
1111 1111 1111 1101	−3
1111 0000 0000 0000	−4,096
1000 0000 0000 0000	−32,768

Fig. 3–4. Two's complement range for sixteen bits.

"It looks to me like negative numbers from −1 through −32,768 could be held using this scheme. But why such a complicated scheme?"

"To make it easier in hardware, friend," said Bob. "This method means that all numbers can be added or subtracted without first testing the sign of each operand. We just go ahead and add or subtract the numbers, and the result will have the correct sign. Suppose that we have the two numbers 0011 0101 0111 0100 and 1011 1111 0000 0000. These are +13,684 and −16,640, respectively. We add them as follows."

```
0011  0101  0111  0100       (+13,684)
1011  1111  0000  0000       (−16,640)
1111  0100  0111  0100       ( −2,956)
```

"The result is 1111 0100 0111 0100, or −2956, just as it should be!"

"That's amazing," said the stranger. "I will study this, sell my binacus, and return to my native land with a fortune!"

"Where might that be?" asked Big Ed.

"Brooklyn," said the stranger. "Goodbye, and thank you. Someday, you'll be able to say that you knew Michael O'Donahue!"

ADDING AND SUBTRACTING BINARY NUMBERS

Before we look at the binacus scheme of representing signed numbers (which duplicates the scheme used on all current microcomputers), let's look at the topic of adding and subtracting binary numbers in general.

Adding binary numbers is much easier than adding decimal numbers. Can you remember when you had to memorize your addition tables? Four and five make nine, four and six make ten, and so forth. In binary, we also have addition tables, but they are much simpler: 0 and 0 make 0, 0 and 1 make 1, 1 and 0 make 1, 1 and 1 make 0, with a carry to the next higher bit position. If the next higher bit position has a 1 and 1, then the table has a fifth entry of 1 and 1 and 1 make 1, with a carry to the next higher bit position. The table is summarized in Fig. 3–5.

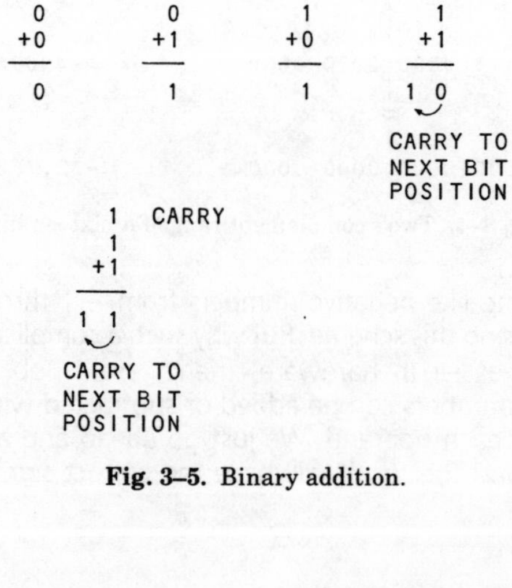

```
   0          0          1          1
  +0         +1         +0         +1
  ──         ──         ──         ──
   0          1          1         1 0
                                   ‿
                               CARRY TO
                               NEXT BIT
                               POSITION


        1  CARRY
        1
       +1
       ──
       1 1
        ‿
     CARRY TO
     NEXT BIT
     POSITION
```

Fig. 3–5. Binary addition.

Let's try this with an example of two unsigned 8-bit numbers, the same ones we've been working with in previous chapters. Suppose we add 0011 0101 and 0011 0111 (53 and 55, respectively).

```
11   111      (Carries)
0011 0101     (53)
0011 0111     (55)
0110 1100     (108)
```

The 1s above the operands represent the carries to the next higher bit position. The result is 0110 1100, or 108, as we had hoped.

Subtraction is just about as easy. 0 from 0 is zero. 0 from 1 is 1. 1 from 1 is zero. And the toughie, 1 from 0, is 1 with a **borrow** from the next higher bit position, just as we borrow in decimal arithmetic. This table is summarized in Fig. 3–6.

```
  0        1        1        0
 -0       -0       -1       -1
 ___      ___      ___      ___
  0        1        0       -1 1
                              \_/
                      BORROW FROM
                      NEXT BIT
                      POSITION
```

Fig. 3–6. Binary subtraction.

Let's try it on some actual numbers. Subtract 0001 0001 from 0011 1010, or 17 from 58:

```
        1           (Borrow)
0011 1010           (58)
0001 0001          −(17)
0010 1001           (41)
```

The one above the operand represents the borrow from the next higher bit position. The answer is 0010 1001, a 41, as we had expected.

Now let's try another example. Subtract 0011 1111 from 0010 1010, a subtract of 63 from 42:

```
1111 111            (Borrows)
0010 1010           (42)
0011 1111          −(63)
1110 1011           (??)
```

The result is 1110 1011, or 235, a result we didn't expect. But wait, could it be? Is it possible? If we apply the rules of two's complement representation and consider 1110 1011 as a negative number, then we have 0001 0100 after changing all ones to zeroes and all zeroes to ones. We then add one to get 0001 0101. The result is −21 when we add the negative sign, which is exactly right. It seems as though we're forced into this two's complement business whether we want to be or not!

TWO'S COMPLEMENT REPRESENTATION

We've pretty well covered all aspects of two's complement notation. If the leftmost bit of an 8- or 16-bit value is treated as a sign bit, then it is either a 0 (positive) or 1 (negative). The actual two's complement representation can then be found by applying the rules discussed earlier—looking at the sign bit, changing all ones to zeroes, all zeroes to ones, and adding one.

Numbers are stored in two's complement fashion either as 8-bit two's complement numbers or as 16-bit two's complement numbers. The sign bit is always in the leftmost bit position, and it is always set (1) for a negative number. The formats for both 8- and 16-bit two's complement representation are shown in Fig. 3–7.

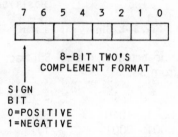

Fig. 3–7. Two's complement formats.

The 8-bit values are used for **displacements** in microprocessor instructions to modify the address of a memory operand. The 16-bit values are used as integer variables in BASIC programs. Either of the 8-bit or 16-bit values can, of course, be used by the assembly language or BASIC programmer to represent anything he chooses.

SIGN EXTENSION

It is possible to add or subtract an 8-bit two's complement number and a 16-bit two's complement number. When these operations are done, the sign of the smaller number must be **extended** to the left until the numbers are of equal length. If this is not done, the result will be incorrect. Take the case of adding the 8-bit value of 1111 1111 (−1) to the 16-bit value of 0011 1111 1111 1111 1111 (+16383). If the sign is not extended, we have:

```
0011   1111   1111   1111        (+16,383)
              1111   1111        (−1)
0100   0000   1111   1110        (+16,638)
```

The result here is +16,638—obviously incorrect. If the sign is correctly extended to occupy all of the bit positions to the left, however, we have:

```
0011   1111   1111   1111        (+16,383)
1111   1111   1111   1111        (−1)
0011   1111   1111   1110        (+16,382)
```

which is the correct result.

ADDING AND SUBTRACTING TWO'S COMPLEMENT NUMBERS

All microprocessors used in current microcomputers have an instruction to add two 8-bit signed numbers and another instruction to subtract two 8-bit signed numbers. In addition, some microprocessors enable addition and subtraction of two 16-bit signed numbers.

The two operands that are involved in the addition or subtraction operation may be any configuration—two positive numbers, a positive and negative number, or two negative numbers. When a subtraction is done, it is convenient to look on the operation as follows. The number to be subtracted (subtrahend) is **negated** by taking its two's complement—changing all ones to zeroes, all zeroes to ones, and, then, adding a one. After this is done, the subtraction is identical to an addition of the negated number. The following is an example. Suppose that 1111 1110 (−2) is to be subtracted from 0111 0000 (+112). First the subtrahend of −2 is negated. The two's complement of 1111

1110 is taken to produce 0000 0010 (+2). Then, the addition is done.

$$
\begin{array}{ll}
0111\ 0000 & (+112) \\
0000\ 0010 & (+2) \\
\hline
0111\ 0010 & (+114)
\end{array}
$$

Bear in mind that the microprocessor does not perform this negate operation before the subtraction takes place. It is simply a convenient way to visualize what happens in the subtraction and it has some bearing on the operations we'll be talking about in the next chapter.

Before we discuss some of the idiosyncrasies of addition and subtraction, such as overflow and carries, and the **flags** of typical microprocessors, try your hand on a few of the following exercises to sharpen your skills in two's complement addition and subtraction.

EXERCISES

1. Add the unsigned numbers below. Show the results in both binary and decimal.

$$
\begin{array}{lll}
01 & 0111 & 010101 \\
11 & 1111 & 101010 \\
\hline
\end{array}
$$

2. Subtract the unsigned numbers below. Show the results both in binary and in decimal.

$$
\begin{array}{lll}
10 & 0111 & 1100 \\
01 & 0101 & 0001 \\
\hline
\end{array}
$$

3. Take the two's complement of the following signed numbers, if necessary, to find the decimal numbers represented:

01101111, 10101010, 10000000

4. What is the 8-bit two's complement form of -1, -2, -3, -30, $+5$, and $+127$?

5. Sign extend the following 8-bit numbers to 16 bits. Show the numbers represented both before and after.

01111111, 10000000, 10101010

6. Add -5 to -300. (No, no—in binary!)

7. Subtract -5 from -300 using binary.

CHAPTER **4**

Carries, Overflow, and Flags

In this chapter, we'll expand some of the concepts presented in the previous chapter. There are some subtle, and some not too subtle, points to consider in working with signed numbers. We'll look at some of them here.

THIS RESTAURANT HOLDS +127 PEOPLE. AVOID OVERFLOW!

Big Ed was just sitting down to a cup of Big Ed's Famous Java after the engineering crowd had left when the restaurant door swung open and a uniformed man walked in.

"Yes, sir, what can I get you?"

"Make it a cuppa coffee and a Danish," the customer answered.

"I notice you're in uniform. Are you in the Air Force from Moffett Field?" asked Ed.

"Naw, I'm a plumber. I came to this area to answer an advertisement in the help wanted section about an 'Overflow Specialist' at Inlog. I figured I fit the bill, but the personnel man just laughed and showed me the door!"

"I think I see the problem, sir," said a slightly built customer who appeared to be about fourteen. A Z-80 reference card stuck out of his shirt pocket. "I'm a programmer who has had a lot of experience with 'overflow.' "

"I didn't think you guys worked wit' pipes."

"Well, we're getting into pipeline processors, but we do work with overflow quite a bit. Mind if I explain?" The plumber raised his eyebrows and waited.

. . . Several cups of Ed's coffee later, the programmer had covered the binary system and simple arithmetic operations.

"So you see, you can hold any value from -128 to $+127$ in eight bits or any value from $-32,768$ to $+32,767$ in sixteen bits. Understand?"

The plumber said, "Right, except for unsigned numbers, in which case the range is 0 through 255, or 0 through 65,535, or $2\uparrow(N-1)$, where N is the size of the register in bits."

"My, you are a fast learner," said the programmer. "Now, to continue. . . . If we perform an addition or subtraction in which the result is greater than $+32,767$ or less than $-32,768$ (or is greater than $+127$ or less than -128), what happens?"

"Well, I guess you get an incorrect result," the plumber said.

"That's true, you definitely get an incorrect result. The result is too great in a positive sense or too great in a negative sense to be held in eight or sixteen bits. In this case, you get **overflow,** because the result overflows the value that can be held in eight or sixteen bits. However, we've designed an overflow **flag** as part of the microprocessor. This flag can be examined by an assembly language programmer to see if overflow occurred after an addition or subtraction."

"Not only do we have an overflow flag, but we have a carry flag, a sign flag, a zero flag. . . ."

"Wait a minute, you mean that guy in personnel wanted a computer engineer that specialized in microprocessor arithmetic?" the plumber said with a grin. "I guess the joke's on me! My PhD is in physics!"

"You have a PhD in physics?" sputtered Big Ed, as most of a mouthful of Big Ed's Java went spewing out across the restaurant floor.

"Yes, I'm only in plumbing to make a living," said the plumber as he walked out the door, shaking his head at the afternoon's embarrassment.

OVERFLOW

As we saw in the preceding example, overflow is possible at any time that an addition or subtraction is done and the result is too large to be put into the number of bits allocated for the result. In the case of most current microcomputers, this would mean when the result is larger than could be held in eight or sixteen bits.

Suppose that we are working with an 8-bit register in the microprocessor. A typical add instruction would add the contents of an 8-bit register, called the **accumulator,** with the contents of another register or memory location. Both **operands** would be signed two's complement operands. What conditions would cause overflow? Any result that was larger than $+127$ or more negative than -128 would produce an overflow. Examples are:

```
  1111  0000  (−16)          0111  1111  (+127)
+ 1000  1100  (−116)         0100  0000  (+64)
  0111  1100  (+124 no!)     1011  1111  (−80 no!)
```

Notice that in both overflow cases, the result was obviously wrong as the sign was opposite from its true sense. Overflow may only occur when the operation is adding two positive numbers, adding two negative numbers, subtracting a negative number from a positive number, or subtracting a positive number from a negative number.

The result of an addition or subtraction in most microprocessors either sets (1) or resets (0) a **flag** bit in the microprocessor. This flag bit can be tested or used in a **conditional jump**—a jump **if** overflow, or a jump **if not** overflow. The testing is done at a machine-language level only, and not in BASIC.

CARRY

Another condition that is important is a **carry** condition. We've seen how carries are used in additions of binary numbers, and how **borrows** (a form of carry) are used in subtractions. The carry that is produced by an addition or subtraction is the carry that is **propagated** off the most significant bit of the result. Here's an example. Suppose that we add the two numbers below:

```
1  1111  111      (Carries)

   0111  1111     (+127)
   1111  1111     (−1)
   0111  1110     (+126)
```

The ones above the operands represent the carries to the next bit position. The leftmost carry, however, "falls off the end" into the bit bucket, an imaginary container. Actually, it doesn't fall into the bit bucket, it normally sets the carry flag in the microprocessor involved. Is this carry useful? Marginally, in this example. There will be a carry as long as the result does not go negative. When the result goes from 0000 0000 to 1111 1111, no carry is produced. As the carry flag may be tested by a conditional "jump if carry" or "jump if no carry" on a machine-language level, the state of the carry is sometimes useful. It is also used for results in shifting, which we'll cover in a later chapter.

OTHER FLAGS

The results of arithmetic operations, such as addition and subtraction, generally set two other flags in the microprocessor. One flag is the "zero" flag. It is set when the result is zero, and reset when the result is not zero. It would be set (1) for an addition of -23 and $+23$, and reset (0) for an addition of -23 and $+22$. As we are constantly comparing values in machine-language programs for the microprocessor, the zero flag is very handy indeed.

Another flag that is generally used is the "sign" flag. The sign flag is set or reset according to the result of the operation. It echoes the bit value in the most significant bit position.

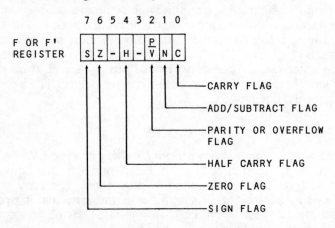

Fig. 4–1. Flags in Z-80 microcomputers.

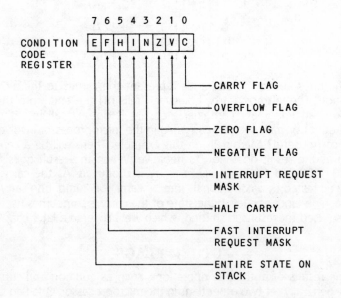

Fig. 4–2. Flags in the 6809E microprocessor.

Both the zero flag and the sign flag can be tested in machine-language programs by conditional jumps, such as "jump if zero," "jump if nonzero," "jump if positive," and "jump if negative." Note that a condition of positive includes the case where the result is zero. Zero is a positive number in two's complement notation.

FLAGS IN MICROCOMPUTERS

The flags in the Z-80 microprocessor are representative of the flags in all microprocessors. They are shown in Fig. 4–1. The flags in the 6809E microprocessor are a second example of flags in a current microprocessor, and are shown in Fig. 4–2.

In the next section, we'll examine logical and shifting operations, both of which also affect the flags, in detail. Before we start that chapter, however, here are some exercises in the use of overflow, carries, and flags.

EXERCISES

1. Which of the following operations will result in overflow? Show all values.

$$
\begin{array}{ccc}
01111111 & 01111111 & 10101111 \\
+00000001 & -00000001 & +11111111 \\
\end{array}
$$

2. Which of the following operations will produce a carry "off the end" that sets the carry flag?

$$
\begin{array}{cc}
01111111 & 11111111 \\
+00000001 & +00000001 \\
\end{array}
$$

3. Suppose that we have a Z(ero) flag and an S(ign) flag in the microprocessor. What are the "states" (0 for reset, 1 for set) of the Z and S flags after the following operations:

$$
\begin{array}{cc}
01111111 & 11011111 \\
+10000001 & -10101010 \\
\end{array}
$$

CHAPTER 5

Logical Operations and Shifting

Logical operations in microcomputers are used to manipulate data on a bit or **field** basis. Shifting also may be used to process bits in binary numbers, or to implement simple multiplication or division.

THE BRITISH ENIGMA

Things were slow at Big Ed's Restaurant in Silicon Valley. Ed was just getting ready to open the local newspaper when he heard a screeching of brakes, and a large red double-decker bus pulled up in front. The doors of the bus opened and a dozen or more people tumbled out. All seemed dressed in tweeds, bowler hats, and other British attire. One even carried a black bumbershoot and kept peering anxiously at the sky.

"Hello, there. Can I help you?"

"Ratherrr. Can you accommodate a group of 17 British computer engineers and scientists?"

"I think so," said Big Ed. "If you don't mind two tables close together, we can put eight of you at Table A and eight of you at Table B. One of you will have to sit in a chair next to Table A. It's my daughter Carrie's chair, but I think it'll do. That'll accumulate . . . , er . . . , accommodate all of you."

"Jolly good," said the spokesman, as the group filled up the two tables and the extra chair (Fig. 5–1).

"Boy, I never realized there were so many computer scientists in Britain," said Ed.

"Just like you Yanks," said the spokesman. "Haven't you heard of Turing, the Colossus Project to break Enigma, or Williams Tubes?"

Big Ed shook his head. "Sorry, no slight intended."

Fig. 5–1. The seating arrangement.

"That's all right, Yank. Let's see, we've got to have lunch and, then, attend a special conference on microprocessors for cricket applications. Gentlemen, may I have your attention, please?

"As you know, the ministry has given us limited funds for this trip. Therefore, we must put certain constraints on lunch today. I've been looking over the menu, and reached the following conclusions.

"One. You may have either coffee (ugh) OR tea, but NOT both.

"Two. You may have soup OR salad or both.

"Three. You may have a sandwich OR a businessman's lunch OR Big Ed's Surprise.

"Four. If you have dessert AND an after-lunch drink, you must pay the additional charge. Any questions?"

One of the older scientists spoke. "Let me see. As I understand it, we have an Exclusive-OR of the coffee and tea, an Inclusive-OR of the soup and salad, an Inclusive-OR of the luncheon, and an AND of the dessert and drink for the additional charge. Correct?"

"That's correct, Geoffrey. All right chaps, chow down!"

"I say, this air conditioner is giving me quite a stiff neck," said one of the young engineers.

"What we must do, then, is to rotate Table A through the extra chair, so that everyone gets a turn at the cool air," said the spokesman. "Every time I say rotate, we'll rotate one chair position."

"Is it really necessary to go through my chair?" asked the engineer seated in the extra chair.

"Well, we have the option of rotating through the chair or not, but I think we'd best do it to be fair."

At regularly spaced intervals, as the lunch progressed, the spokesman would shout out "Rotate!" and the group at Table A and the extra chair would move as shown in Fig. 5–2. Finally the entire group at Table A got up, paid their bills, and left.

"I think we're just about done here, sir," said the group at Table B.

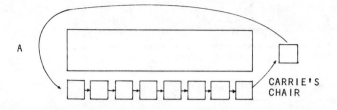

Fig. 5–2. The group rotates.

"All right, then, shift left logically into the empty chair so I can look at your bill," said the spokesman.

The scientist nearest to the cash register got up and slid into the empty chair. The spokesman examined the bill. "Next shift!" he shouted. The person in the extra chair went to the cash register, and the next person slipped into the extra chair. This process continued until Table B was empty (see Fig. 5–3).

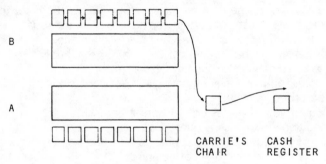

Fig. 5–3. The group shifts.

"Jolly good lunch," said the spokesman to Big Ed as he left.

"It was nice having you," said Ed. "Hope your cricket seminar goes well."

"We're not too worried about that. It's the arachnid problem that concerns us," said the spokesman as he unfurled his umbrella and strode out the door into the sunny San Jose skies. "Cheerio!"

LOGICAL OPERATIONS

All microcomputers have the ability to perform the **logical** operations of ANDing, ORing, and Exclusive-ORing on a machine-language level. Also, most versions of BASIC allow ANDing and ORing.

All logical operations operate on a "bit position" basis. There are no carries to other bit positions. On a machine-language level, logical operations operate on one byte of data; BASIC allows the logical op-

```
       0            0            1            1
  OR   0       OR   1       OR   0       OR   1
       ─            ─            ─            ─
       0            1            1            1        RESULT


          00110111  ⎫
  OR      01011011  ⎬  8-BIT OR
          ────────  ⎭
          01111111
```

Fig. 5–4. OR operation.

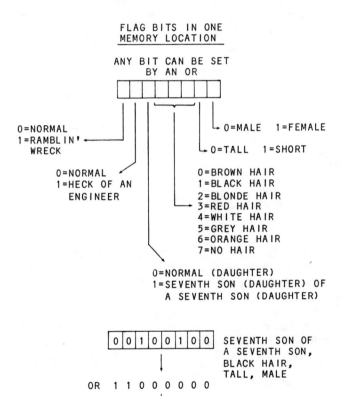

Fig. 5–5. Using an OR operation.

erations to take place with two-byte operands. There are always two operands involved, and one result.

ORs

The OR operation is shown in Fig. 5–4. Its **truth table** states that there will be a 1 bit in the result if either operand has a 1 bit or both operands have a 1 bit.

The OR operation is used on a machine-language level to **set** a bit in a value. A typical application might use the eight bits of a memory location as eight **flags** for various conditions, as shown in Fig. 5–5. The OR would not be as widely used in BASIC, but it comes up occasionally, as for example, when setting the bits of a video-display byte to lower case, as shown in Fig. 5–6.

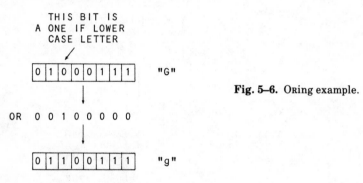

Fig. 5–6. ORing example.

ANDs

The AND operation is shown in Fig. 5–7. It also operates only on a bit-wise level with no carries to other bit positions. The result of the AND is a one if both operands have a 1 bit; if either has a 0, the result is zero.

The AND is used on a machine-language level primarily to **reset** a bit, or to **mask out** certain parts of an 8-bit word, as shown in Fig. 5–8. In BASIC, the AND operation would have more limited applica-

```
        0            0            1            1
AND     0      AND   1      AND   0      AND   1
       ---          ---          ---          ---
        0            0            0            1
```

```
          00110111 ⎤
AND       01011011 ⎬  8-BIT AND
          --------
          00010011 ⎦
```

Fig. 5–7. AND operation.

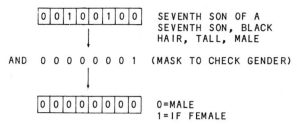

Fig. 5–8. ANDing example.

tions. An example is shown in Fig. 5–9; it checks for multiples of 32 for a line count of 32 lines per page.

Exclusive-ORs

The Exclusive-OR is shown in Fig. 5–10. Its rules state that the result is a 1 if one bit OR the other bit, but not both, are 1s. In other words, if both bits are 1s, the result is 0.

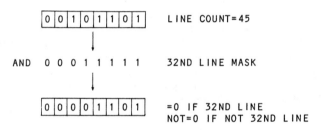

Fig. 5–9. Another ANDing example.

```
       0            0            1            1
XOR    0     XOR    1     XOR    0     XOR    1
      ___          ___          ___          ___
       0            1            1            0       RESULT
```

```
       00110111  ⎫
       01011011  ⎬  8-BIT XOR
       _____
       01101100  ⎭
```

Fig. 5–10. Exclusive-OR (XOR) operation.

The Exclusive-OR is used infrequently in both machine language and in BASIC. One application is shown in Fig. 5–11 where the least significant bit is used as a "toggle" bit to indicate the number of the pass—odd or even.

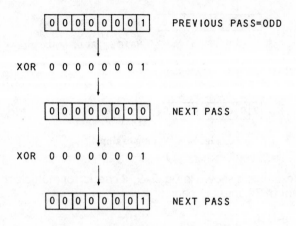

Fig. 5–11. XOR example.

Other Logical Operations

There are a number of other logical operations that are implemented in BASIC and in machine language. One of these is the NOT operation. NOT is similar to the **negate** operation discussed in Chapter 4, except that it forms the **one's complement.** The one's complement of a number is formed by changing all the ones to zeroes and all the zeroes to ones, **without** adding a one. What effect does this have? Let's look at an example of a signed number.

Suppose that we have the value 0101 0101. If we NOT the number, we get 1010 1010. The original value was +85. The result is a negative number which, when reconverted by the two's complement rules, comes out to 0101 0101 + 1 = 0101 0110 = −86. You might say, therefore, that the NOT adds one to the number and, then, negates it. The machine-language version of the BASIC NOT operation is normally called CPL—for (One's) Complement.

The BASIC NOT operation can be used to test logical conditions in a BASIC program, as shown in Fig. 5–12. The machine-language CPL is used somewhat infrequently.

```
1010    IF NOT (PRINTER) THEN

        PRINT "NO PRINTER-BUY ONE-I'LL WAIT"

        ELSE LPRINT "RESULT=";A
```

Fig. 5–12. NOT operation.

SHIFT OPERATIONS

The British restaurant party illustrated two types of shifts commonly used on microcomputers—**rotates** and **logical** shifts. These are available on a machine-language level but not in BASIC, and generally operate on eight bits of data. They are tied in with the carry flag discussed in the last chapter.

Rotates

A rotate shift was first illustrated in Fig. 5–2 with Table A in the restaurant. Now, let's look at Fig. 5–13, where the data is rotated right or left, one bit position at a time. Although larger computers will allow any number of shifts with one instruction, current microcomputers allow only one shift with each instruction.

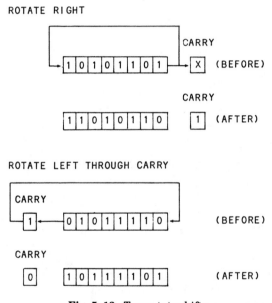

Fig. 5–13. Two rotate shifts.

As the data is rotated out the end of the microprocessor register or memory location, it either goes back into the opposite end of the register or memory location, or goes into the carry flag. If the data goes through the carry, it is really a **9-bit rotate.** If the data bypasses the carry, it is an **8-bit rotate.** *In either case, the bit shifted out always goes into the carry,* as shown in Fig. 5–13.

The carry flag can be tested by a conditional jump instruction on a machine-language level to effectively test whether a one bit or zero bit was shifted out. The rotate shift is used to test a bit at a time for such operations as multiplication (see next chapter) or the alignment of data for **masking** with an AND.

Logical Shifts

The second type of shift that was illustrated in the restaurant anecdote was a **logical** shift. The logical shift is not a rotate. Data falls off the end into the bit bucket, just as the scientist left the restaurant. As each bit is shifted out, though, it does go into the carry so that the carry always holds the result of the shift. The opposite end of the register or memory location is filled with zeroes as each bit is shifted. Here, again, one bit position at a time is shifted. The logical shift operation is shown in Fig. 5–14.

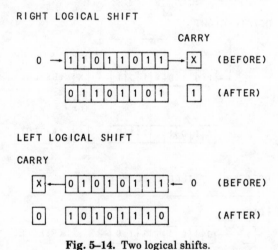

Fig. 5–14. Two logical shifts.

We couldn't really say much about the rotate in regard to what happened to the contents arithmetically. This was because the data recirculated back into the register and, in an arithmetic sense, the results were not predictable.

In the case of the logical shift right or the logical shift left, however, the results are very predictable. Let's look at some examples. Suppose that we have the value 0111 1111 with the carry set to some value. We'll express the register and carry by nine bits with the carry on the right, as in 0111 1111 x. After a shift-right logical operation, we have 0011 1111 1. The original value was + 127. After the shift, the value is + 63 with the carry set. It appears that a logical right shift divides by two and puts the remainder of 0 or 1 into the carry! This is true, and the shift-right logical operation can be used any time that a divide by 2, 4, 8, 16, or other power of two is called for.

How about a shift left? Absolutely right! A shift-left logical operation multiplies by 2. As an example, consider x 0001 1111, where x is the state of the carry flag. After a logical shift left, the result is 0 0011 1110. The original number was +31 and the result is +62, with the carry reset by the most significant bit. The logical shift left can be used any time that a number is to be multiplied by 2, 4, 8, or any other power of two.

Arithmetic Shifts

There is a problem that arises when a logical shift is used with signed numbers. Consider the case of the number 1100 1111. This is a value of −49. When the number is shifted right in a logical shift, the result is 0110 0111, representing the value +103. Obviously, the shift did not divide the signed number of −49 by 2 to produce a result of −24.

To solve this problem of shifting arithmetic data, an arithmetic shift is often implemented in microcomputers. The arithmetic shift **extends**

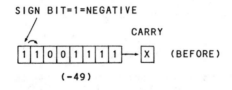

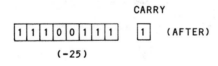

Fig. 5–15. Arithmetic right shift.

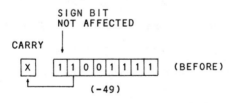

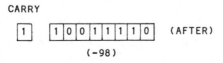

Fig. 5–16. Arithmetic left shift.

the sign as it shifts right, so that the shift is (almost) arithmetically correct. If an arithmetic shift were used in the preceding example, the result would be as shown in Fig. 5–15.

What about arithmetic left shifts? In some microcomputers, the arithmetic left shift retains the sign bit and shifts the next most significant bit out of the register and into the carry, as shown in Fig. 5–16. In other microcomputers, there is no true left shift.

In the next chapter, we'll see how shifting can be used to implement many different types of multiplication and division **algorithms.** In the meantime, try your hand with the following test questions.

EXERCISES

1. OR the following sets of 8-bit binary operands.

$$
\begin{array}{cc}
10101010 & 10110111 \\
\text{OR } 00001111 & \text{OR } 01100000
\end{array}
$$

2. Exclusive-OR the following sets of 8-bit binary operands.

$$
\begin{array}{cc}
10101010 & 10110111 \\
\text{XOR } 00001111 & \text{XOR } 01100000
\end{array}
$$

3. Bits 3 and 4, of a location in memory, hold a code as follows: 00 = BROWN HAIR, 01 = BLACK HAIR, 10 = BLONDE HAIR, 11 = NO HAIR. Using an AND operation, show how these bits could be put into an 8-bit result alone. The location holds XXXYYXXX, where X = unknown bit and Y = code bit.

4. Negate the following signed operands. Show their decimal equivalents:

$$00011111, \quad 0101, \quad 10101010$$

5. Perform a left rotate on these operands:

$$00101111, \quad 10000000$$

6. Perform a right rotate on these operands:

$$00101111, \quad 10000000$$

7. Perform a left rotate with carry on these operands and carry:

$$C = 1 \quad 00101111, \quad C = 0 \quad 10000000$$

8. Perform a right rotate with carry on these operands and carry:

$$00101111 \quad C = 0, \quad 10000000 \quad C = 1$$

9. Perform a right logical shift of the following operands. Show the carry after the shift and the decimal values of the operands before and after the shifts:

$$01111111, \quad 01011010, \quad 10000101, \quad 10000000$$

10. Perform a left logical shift of the following operands. Show the carry after the shift and the decimal values of the operands before and after the shifts:

01111111, 01011010, 10000101, 10000000

11. Perform an arithmetic right shift on the following operands and show the decimal values before and after the shift:

01111111, 10000101, 10000000

CHAPTER **6**

Multiplication and Division

Most of today's microcomputers do not have "built-in" multiplication and division instructions. As a result, multiplication and division have to be done in "software" routines, at least in machine language. In this chapter, we'll look at some of the ways that multiplication and division can be implemented in software.

ZELDA LEARNS HOW TO SHIFT FOR HERSELF

"Hi, Don. How was the lunch?" asked Zelda, the serving person, of an engineer from Inlog who had just appeared at the cash register.

"Fine, just fine. Well, the meat was a little tough. . . ."

"That always happens when it sits around for so long," said Zelda, as she took the bill. "Let's see, a cup of coffee . . . one leftover meatloaf sandwich . . . and twelve lo-cal desserts. . . ." As Zelda read off each item she performed some type of operation out of sight beneath the counter. On the last item, twelve lo-cal desserts, she spent a great deal of time.

"Zelda, what are you doing down there?" asked the engineer.

"Well, Don, Big Ed wants us to get used to working in binary, what with this restaurant being in the center of microcomputer manufacturers and all. He wants us to figure out all of the checks in binary to give us practice. I'm okay when it comes to adding and subtracting in binary, but multiplying gives me some trouble."

"What method are you using, Zelda? Maybe I can help."

"Every time I multiply, I use successive addition. Like in this case, where you had the twelve lo-cal desserts. Now, they cost 65 cents each, so I add 1100, which is twelve in binary, 65 times."

"Well that certainly will work, all right," said Don, as he suppressed a giggle. "That successive addition method is accurate, but it takes a good deal of time. Let me show you a faster method; one called shift and add."

He quickly snapped the tablecloth off of a nearby table, leaving the silverware and accessories relatively unmolested. He drew out an engineer's mechanical pencil. "Pity we don't have any quadrille paper, but these small checks in the pattern will have to do."

"Let's take that 65 cents per lo-cal dessert for twelve items example. First of all, we'll draw two registers. The **partial product** register is sixteen bits wide like so." (See Fig. 6–1.) "The **multiplicand** register is eight bits wide like so. Now, into the upper eight bits of the partial product register, we'll put the twelve, the **multiplier.** We've **padded** it out to eight bits to make it 00001100. The remainder of the register we'll **zero out.** Next, we'll put the multiplicand into the multiplicand register.

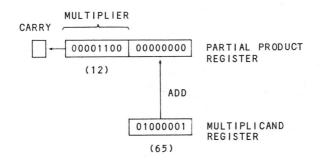

Fig. 6–1. Shift and add multiplication.

"Now, we're all set to do the multiplication. We'll have eight steps for this. We computer engineers call them **iterations.** For each iteration, we'll shift the multiplier one bit position to the left. The bit shifted out will go into the carry. If the bit in the carry is a 1, then we'll add the multiplicand to the partial product register. If the bit in the carry is a 0, we won't do the addition. At the end of eight iterations, we'll be done."

"But, won't the add to the partial product register **clobber** the multiplier in the upper eight bits?" asked Zelda, obviously proud that she had picked up some computer jargon.

"No, it won't. Remember that data is being shifted out to make room for the possibly expanding partial product. After eight iterations, it will have shifted out entirely, and the partial product register will now have the final product of the multiplication. Look, I drew out all the iterations for this case." (See Fig. 6–2.)

"Oh, yeah. Gee thanks, Don. I'll use this for sure. Now, let me figure out the rest of your bill. The total is 15.63 plus sales tax. That's 1563 cents divided by 100 times six. Let's see 01100100 from 011000011011 leaves 010110110111—that's once. 01100100 from 010110110111 leaves 010101010011—that's twice. 01100100 from. . . ."

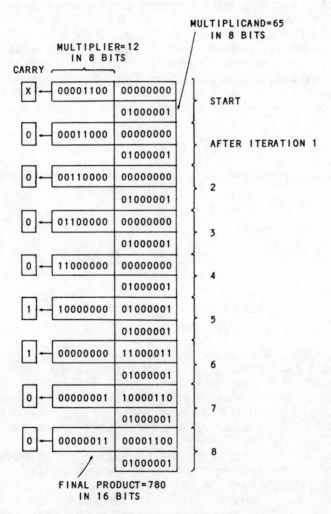

FINAL PRODUCT=780
IN 16 BITS

Fig. 6–2. Shift and add multiplication example.

MULTIPLICATION ALGORITHMS

Successive Addition

Zelda was using a straightforward method of multiplication called **successive addition** (Fig. 6–3). In it, the multiplicand (the number to be multiplied) is multiplied by the multiplier. The process is done by zeroing a result, called the partial product, and adding the multiplicand to the partial product for the number of times that is equal to the multiplier. The preceding example was 65 times 12, which Zelda solved by adding 12 to the partial product 65 times.

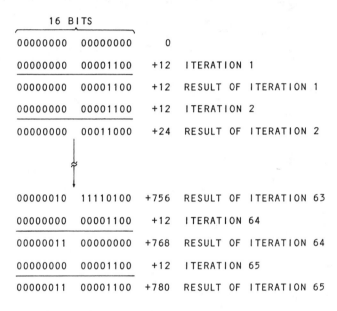

```
      16 BITS
 ┌─────────────────────┐
 00000000  00000000      0
 00000000  00001100     +12    ITERATION 1
 00000000  00001100     +12    RESULT OF ITERATION 1
 00000000  00001100     +12    ITERATION 2
 00000000  00011000     +24    RESULT OF ITERATION 2

                ≈

 00000010  11110100    +756    RESULT OF ITERATION 63
 00000000  00001100     +12    ITERATION 64
 00000011  00000000    +768    RESULT OF ITERATION 64
 00000000  00001100     +12    ITERATION 65
 00000011  00001100    +780    RESULT OF ITERATION 65
```

Fig. 6–3. Successive addition multiplication example.

Although this method is straightforward, it is **very** time-consuming in the average case. Suppose that we're working with an "eight by eight" multiplication. An 8-bit by 8-bit multiplication produces a 16-bit product, maximum. The average multiplier is 127, if an **unsigned multiplication** is done. This means that in the average case, 127 separate additions to the partial product would have to be done.

Contrast this with Don's **shift and add** technique of eight iterations! In the worst case, the successive addition will involve 255 additions and, in the best case, 1 addition. The successive addition multiplication is best left for cases where the multiplier is fixed at some low constant value, below 15 or so, or where infrequent multiplications are required.

Successive Addition of Powers of Two

The powers of two successive addition method is a multiplication method that is often used when the multiplier will be a fixed value. Suppose that we must always multiply an amount by ten. Ten can be broken up into a number of sums. One combination of these is 8 + 2.

As we saw in the last chapter, logically shifting to the left multiplies a value by two. To multiply by ten, a combination of shifts and additions can be done to effect the multiply. The operation goes like this:

1. Shift the multiplicand by two to get two times X.
2. Save this value as "TWOX."
3. Shift the multiplicand by two to get four times X.
4. Shift the multiplicand by two to get eight times X.
5. Add in "TWOX" to the multiplicand to get ten times X.

This procedure is shown in Fig. 6–4, for a multiplier of 10. It can be used any time the multiplier is fixed. Multiplying by 35, for example, could be done by five shifts of the multiplicand to get 32X, and an addition of 2X and 1X − 32 + 2 + 1 = 35.

```
MULTIPLY 11 BY 10

         00001011   MULTIPLICAND=11

         00001010   MULTIPLIER=10

SHIFT MULTIPLICAND BY TWO (TWOX)

┌────────── 00010110   MULTIPLICAND=11X2=22.SAVE

│   SHIFT MULTIPLICAND BY TWO (FOURX)

│        00101100   MULTIPLICAND=22X2=44

│   SHIFT MULTIPLICAND BY TWO (EIGHTX)

│ ADD    01011000   MULTIPLICAND=44X2=88
└──────→ 00010110   PREVIOUS SHIFT
         01101110   PRODUCT=110
```

Fig. 6–4. Successive addition example using powers of two.

Shift and Add Multiplication

The shift and add multiplication that Don showed Zelda is the most commonly used method for computers that do not have a built-in "hardware" multiplication capability. It emulates a paper and pencil technique that closely follows the normal decimal multiplication method that we would use. For example, Fig. 6–5 shows a paper and pencil binary multiplication that is similar to a decimal multiplication. The only real difference between the paper and pencil binary method

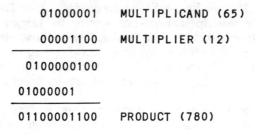

```
01000001    MULTIPLICAND (65)

00001100    MULTIPLIER (12)
_____
0100000100

01000001
_____
01100001100   PRODUCT (780)
```

Fig. 6–5. Paper and pencil binary multiplication.

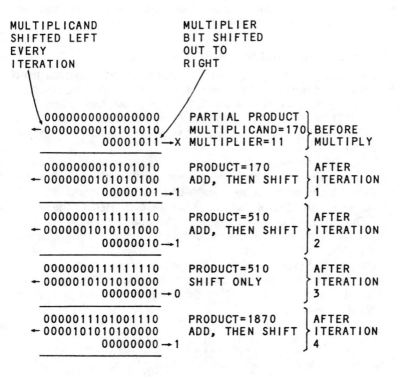

```
MULTIPLICAND              MULTIPLIER
SHIFTED LEFT              BIT SHIFTED
EVERY                     OUT TO
ITERATION                 RIGHT


 0000000000000000    PARTIAL PRODUCT  ⎤
-0000000010101010    MULTIPLICAND=170 ⎬ BEFORE
        00001011→X   MULTIPLIER=11    ⎦ MULTIPLY
 _____
 0000000010101010    PRODUCT=170      ⎤ AFTER
-0000000101010100    ADD, THEN SHIFT  ⎬ ITERATION
        00000101→1                    ⎦ 1
 _____
 0000000111111110    PRODUCT=510      ⎤ AFTER
-0000001010101000    ADD, THEN SHIFT  ⎬ ITERATION
        00000010→1                    ⎦ 2
 _____
 0000000111111110    PRODUCT=510      ⎤ AFTER
-0000010101010000    SHIFT ONLY       ⎬ ITERATION
        00000001→0                    ⎦ 3
 _____
 0000011101001110    PRODUCT=1870     ⎤ AFTER
-0000101010100000    ADD, THEN SHIFT  ⎬ ITERATION
        00000000→1                    ⎦ 4
 _____
```

Fig. 6–6. Example of the "multiply until zero multiplier" method.

and the shift and add binary method is in the shifting. With paper and pencil, the multiplicand is shifted to the left and then added. With shift and add, the multiplicand is stationary and the partial product is shifted, as shown in Fig. 6–2.

This shift and add technique can be used for any size multiplier and multiplicand. The rule for the size of the product is this: **The product of a binary multiplication can never be larger than the total number of bits in the multiplier and multiplicand.** In other words, if the multiplier and multiplicand are eight bits and eight bits, allow sixteen bits for the product. If the multiplier and multiplicand are twelve bits and eight bits, allow twenty bits for the product, and so forth.

Another interesting fact about shift and add multiplication is that the shifting can be in reverse. We can work with multiplier bits starting from the **least significant** end! In this case, we can stop the process when the multiplier is zero, which means that we only have to perform as many iterations as there are **significant bits** in the multiplier. This reduces the average time of multiplication to one-half the maximum multiplier value. This efficient method is shown in Fig. 6–6.

Signed Vs. Unsigned Multiplication

In all of the preceding examples, we've been considering **unsigned** multipliers and multiplicands. The products formed in these cases are absolute numbers without a sign bit. For example, multiplying 1111 1111 (255) by 1011 1011 (187) produces a product of 1011 1010 0100 0101 (47,685) where the most significant bit is $2\uparrow15$, and not a sign bit.

What about **signed** multiplication, where the multiplier and multiplicand are signed two's complement numbers? In this case, the numbers can be converted to their **absolute values,** the multiplication can proceed, and the product can then be converted to the proper sign. Of course, a plus quantity times a plus quantity is a plus, a minus times a plus is a minus, and a minus times a minus is a plus quantity, just as in decimal arithmetic.

Here's a good example of one of our logical operators, the Exclusive-OR. If an Exclusive-OR is taken of the multiplicand and multiplier, then the most significant (sign) bit of the result will be the sign of the product! For example,

$$
\begin{array}{ll}
1011 \quad 1010 & (-70) \\
\underline{0000 \quad 1010} & (+10) \\
1011 \quad 0000 & (\text{xor})
\end{array}
$$

produces an XOR whose most significant bit is a 1; therefore, the product will be negative.

The algorithm for a signed multiplication is this:

1. XOR the multiplier and multiplicand. Save the result.
2. Take the absolute value of the multiplicand by changing a negative number to positive, if necessary.
3. Take the absolute value of the multiplier by changing a negative number to positive, if necessary.
4. Multiply the two numbers by the standard shift and add method.
5. If the result of the XOR has a 1 bit in the sign, change the product to a negative number by the two's complement method of complementing it (negate operation).

There is one minor problem in the preceding method. Overflow was not possible in the unsigned method, but it is possible in the signed method, for one case only. When both the multiplier and multiplicand are maximum negative values, the product will overflow. For example, if in an "eight by eight" multiply, the multiplier and multiplicand are both 1000 0000 (-256), then the product will be $+65,536$, which is too large to be held in sixteen bits. This condition can be tested before the multiplication takes place.

DIVISION ALGORITHMS

Software division algorithms are somewhat harder to implement than multiplication algorithms. One of the methods of division is successive subtraction, the one that Zelda was using as we left her. A second is a bit-by-bit "restoring division."

Successive Subtraction

Successive subtraction is shown in Fig. 6–7. In this method, the **quotient** is cleared to zero initially. The **divisor** (the operand that "goes into" the **dividend**) is subtracted from the dividend successively until the dividend "goes negative." Each time the divisor is subtracted, the quotient count is incremented by one. When the dividend goes negative, the divisor is **added** once to the **residue** to restore the remainder. (The residue is the amount of dividend remaining.) The count is the quotient of the division while the remainder is in the residue, as shown in Fig. 6–7.

As in the case of successive addition for multiplication, the successive subtraction is very slow for the average case. If we are working with a 16-bit dividend and an 8-bit divisor, the average quotient is 255. A total of 255 subtractions is just as intolerable as in the multiplication case. Successive subtraction for a division is fine where the size of the divisor is large compared to the size of the dividend; for example, if the divisor were 50 and the dividend was a maximum of 255, then only

				COUNT (QUOTIENT)
DIVIDEND	00000110	00011011	1563	0
DIVISOR		01100100	-100	
	00000101	10110111		1
		01100100	-100	
	00000101	01010011		2
		01100100	-100	
	00000100	11101111		3
		01100100	-100	
	00000100	10001011		4
		01100100	-100	
	00000100	00100111		5
		01100100	-100	
	00000011	11000011		6
		01100100	-100	
	00000011	01011111		7
		01100100	-100	
	00000010	11111011		8
		01100100	-100	
	00000010	10010111		9
		01100100	-100	
	00000010	00110011		10
		01100100	-100	
	00000001	11001111		11
		01100100	-100	
	00000001	01101011		12
		01100100	-100	
	00000001	00000111		13
		01100100	-100	
	00000000	10100011		14
		01100100	-100	
	00000000	00111111		15=QUOTIENT
		01100100	-100	
(DIVIDEND	11111111	11011011		
GOES		01100100	+100	(RESTORE
NEGATIVE)				REMAINDER)
	00000000	00111111	REMAINDER=63	

Fig. 6–7. Successive subtraction method of division.

five subtractions would have to be done in the "worst case," which would make this an efficient division.

Restoring Division

The answer to a relatively fast general-purpose division technique lies in a bit-by-bit divide. The restoring division method emulates the way we divide by paper and pencil. Fig. 6–8 shows the scheme. We are dividing a 16-bit dividend with an 8-bit divisor. The general rule for the size of the quotient, by the way, is that it must be equal to the number of bits in the dividend, as the value 1 is a legitimate divisor.

We start with the two absolute values (positive numbers) of +96 for the divisor and +3500 for the dividend. First of all, 0110 0000 (96)

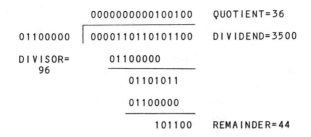

Fig. 6–8. Bit-by-bit division example.

is subtracted from the first 0 of the dividend of 0000 1101 1010 1111. Of course, this subtraction is impossible, and the result is negative. If the result is negative after any subtraction, we "restore" the previous residue by adding back the divisor, and we do so in this case. We continue this subtraction, test, and the restore/nonrestore for each of the sixteen bits in the dividend. Any time the result is negative, we restore by adding back the divisor and put a 0 in the quotient bit. Any time the result is positive, we do not restore and put a 1 bit in the quotient bit. At the end of sixteen iterations, the quotient holds the final value, and the residue is the remainder of the operation. The residue may have to be restored by one final add to obtain the true remainder.

This paper and pencil technique can be almost exactly duplicated in the microcomputer. The dividend is put into a 24-bit register (three 8-bit registers). The divisor is put into an 8-bit register. The two registers are aligned as shown in Fig. 6–9. The divisor is subtracted from the upper eight bits of the dividend. A restore is done by adding back the divisor if necessary.

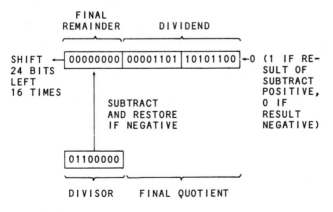

Fig. 6–9. Bit-by-bit division implementation.

After the subtraction and possible restore are done, the dividend is shifted left one bit position. At the same time that the shift occurs, the quotient bit of 0 or 1 is shifted left into the dividend register from the right end. At the end of sixteen iterations, sixteen quotient bits will have been shifted into the dividend register, and they will be in the lower sixteen bits. The upper eight bits will hold an 8-bit remainder, provided that any final restore has been done.

Signed Vs. Unsigned Divisions

All of the preceding divisions were of unsigned operands. As in the case of multiplication, the easiest route for a signed division is to find the "sense" (positive or negative) of the final product, take the absolute value of the dividend and divisor, and then perform an unsigned division, changing the quotient to a negative value if a plus quantity was divided by a minus quantity, or a minus was divided by a plus.

There is one minor problem. When $-32,768$ is divided by -1, the result of $+32,768$ will overflow the quotient. A check can easily be made of this condition before the division.

In the next chapter, we'll consider "multiple-precision" arithmetic operations that will allow us to work with greater values than can be held in sixteen bits. Before we get on to this topic, however, the following exercises are in search of readers who want to sharpen their skills in multiplication and division algorithms and calculations.

EXERCISES

1. What is the result of the unsigned eight-by-eight multiplication of 11111111 and 11111111? What are the equivalent decimal values?

2. What is the result of the unsigned division of 1010101011011101 by 00000000? What is the result of the unsigned division of 1111111111111111 by 00000001? What conditions cannot be allowed in computer division?

3. If all operands are signed numbers, what will be the sign of the result of these operations:

$$10111011 \times 00111000 = ?$$
$$1011011100000000/10000000 = ?$$

CHAPTER **7**

Multiple Precision

Many times it is necessary to work beyond the **precision** offered by eight bits or sixteen bits. Real-world quantities such as physical constants, accounting data, and other numeric values often exceed the $+32,767$ to $-32,768$ range of even 16-bit quantities. Larger values can be expressed by using a number of bytes to hold each operand. In this chapter, we'll see how this can be done by using 8-bit bytes.

IS THE FIBONACCI SERIES RELATED TO ITALIAN SOCCER?

" 'Scusa, me. Is this-a Big Ed's Ristorante?"

Big Ed turned around to see a smiling man with a large notebook just entering his restaurant. "Yes, sir, it is. What can I get you?"

"I'd like-a some lasagne and Chianti, if-a you please."

"Certainly, sir. By the way, may I ask you to dispense with the Italian accent? It makes the writer very nervous. He's not too good with foreign accents . . ."

"Oh, I'm sorry. I have to use it on lecture tours. I guess I've fallen into bad habits."

"Lecture tours?"

"Yes, one of my distant relatives was Leonardo of Pisa, also known as Fibonacci, the great Italian mathematician. I'm carrying on his work. That's why I'm here in Silicon Valley. I want to use a microcomputer to help me analyze the Fibonacci Series. I was a little dismayed to find out that they normally don't have enough precision to do this."

"What do you mean by that? I thought microcomputers could handle any size number."

"Well, they can handle **approximately** any size number by **floating-point,** but that doesn't give you exact precision. For example, the value 122,234,728,956 would probably only be kept in BASIC as 122,235,000,000 with the rest of the digits lost. I need exact precision for my work on the Fibonacci Series."

"I really don't follow soccer. . . ."

"No, you-a see, sorry. . . . You see, the Fibonacci Series goes like this. If you take the numbers 1 and 1 and add them, you get 2. That's the third term in the series. Now take the 1 and 2 and add them to get 3. That's the fourth term. Now add 2 and 3 to get 5. That's the fifth term. Well, if you keep on going like that, you'll have the simple series of 1, 1, 2, 3, 5, 8, 13, 21, 34, 55, and so forth."

"Well, that seems simple enough. Is it good for anything?"

"It has certain applications in mathematical modeling and in nature, but the main interest about the series seems to be with a hard core of people who delight in mathematical games. There are literally tens of thousands of people around the world investigating the properties of the Fibonacci Series. Why, in my Inlog lecture alone, 134 engineers, programmers, and scientists showed up, in addition to one gentleman demanding the lecture hall rental fee!

"To make a long story short, though, my main problem in working with the series is that the terms get very big, very fast. The twenty-fourth term is 46,368, too much for sixteen bits to handle. I had to go to multiple precision to process the larger terms of the series."

"Multiple precision? How does that work?"

"Well, as you probably know from associating with all of the computer manufacturing people around here, eight bits will hold values up to 255 and sixteen bits will hold values up to 65,535, if the numbers are unsigned, of course.

"I wanted to hold numbers up to 18,446,745,073,709,548,616—that would cover all Fibonacci numbers up to almost the one-hundredth term."

"You'd need hundreds of bytes to do that!" exclaimed Ed.

"Not really—four bytes hold 4,294,967,296, six bytes hold 281,-474,976,710,656, and eight bytes hold 18,446,745,073,709,548,616. So you can see that if I had a program capable of dealing with only eight bytes, I'd have more than enough precision. All that program has to do is to work with eight-byte operands for addition and subtraction. I finally found a software house that specialized in programs for computing large numbers in microcomputers. . . ."

"Yes, I'm all ears. . . ."

"Alas, alack, they had gotten a huge federal contract to process federal budget programs. They're now in Washington and I'm here, looking for another consulting company. . . ."

ADDITION AND SUBTRACTION USING MULTIPLE PRECISION

Multiple precision is not generally used to handle large numbers (floating-point, as covered in Chapter 10, is used instead), but it does come in handy for certain types of processing, such as Fibonacci's problem, high-speed operations, or high-precision operations.

In multiple-precision addition and subtraction, two operands that are a number of bytes long are added or subtracted together. What is the appearance of the operands? Suppose we have determined that we want to represent numbers up to a size of one half of 18,446,745,073,709,548,616. We know from the Fibonacci conversation that this magnitude could be held in a signed number of eight bytes, or 64 bits. The value of +9,223,372,536,354,779,313 would be represented by

```
01111111   11111111   11111111   11111111
11111111   11111111   11111111   11111111
```

The bits would be numbered in our standard representation of powers of two, with bit 0 on the right-hand side and bit 63 on the left-hand side as a **sign bit.** The value of −9,223,372,536,354,779,314 (note that this is one more in magnitude than the positive number) would be represented by

```
10000000   00000000   00000000   00000000
00000000   00000000   00000000   00000000
```

Other values would be between these two limits. Note that there is only one sign bit and that the number is treated as a single group of 64 bits, even though it is **physically separated** into eight bytes.

The addition of two operands in this eight-byte precision involves taking the bytes of each of the operands from right (least significant) to left (more significant) and adding them together. Any **carry** from the addition of the last, less significant byte must be added in.

While carries **propagate** to the next bit position inside each byte as a normal result of the addition, any carry to the next bit position affects the next higher byte, and the carry must be saved and added in. The carry flag is used to record any carry from the previous addition, and an "add with carry" instruction is used to add-in the previous carry.

For an eight-byte operand, the first addition is a simple "add" (there is no previous carry), while the next seven additions are "add with carry" instructions. Fig. 7–1 shows this operation for a sample addition of two 8-bit operands.

A multiple-precision subtraction works in much the same way except

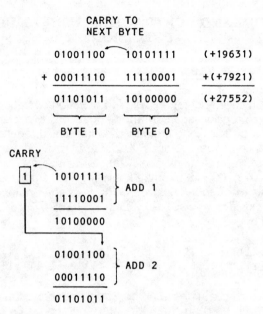

Fig. 7-1. Two-byte multiple-precision addition.

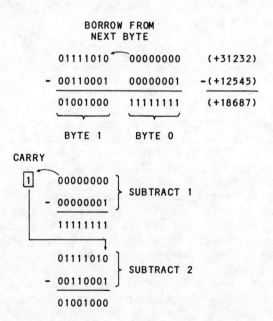

Fig. 7-2. Two-byte multiple-precision subtraction.

that a "borrow" from the next higher byte is used instead of a carry. The first subtraction is a plain subtraction, while the remaining subtractions are "subtract with borrow" operations, commonly called "subtract with carry" operations. Fig. 7–2 shows the subtraction operation for two 8-bit operands.

The preceding addition and subtraction operations work fine with any mixture of two's complement numbers; the operands do not have to be absolute.

When multiple-precision numbers are stored in memory, one cautionary note should be added. In Z-80 microcomputers, the normal memory storage of sixteen bits of data, including memory addresses and two-byte values, is the least significant byte, followed by the most significant byte, as shown in Fig. 7–3. If a multiple-precision number is stored in the most significant byte to the least significant byte fashion,

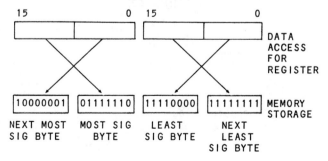

2,122,448,880 IN QUADRUPLE PRECISION

01111110	10000001	11111111	11110000
MOST SIG BYTE	NEXT MOST SIG BYTE	NEXT LEAST SIG BYTE	LEAST SIG BYTE

EIGHT-BIT REGISTER ACCESS:

01111110	10000001	11111111	11110000	MEMORY STORAGE
MOST SIG BYTE	NEXT MOST SIG BYTE	NEXT LEAST SIG BYTE	LEAST SIG BYTE	

SIXTEEN-BIT REGISTER ACCESS:

10000001	01111110	11110000	11111111	MEMORY STORAGE
NEXT MOST SIG BYTE	MOST SIG BYTE	LEAST SIG BYTE	NEXT LEAST SIG BYTE	

Fig. 7–3. Memory storage with multiple precision.

there is no problem if the data is accessed one byte at a time to: (1) load the data into a register, (2) add or subtract the second operand, and, then, (3) store the result byte back into memory. If the data is loaded into a 16-bit register, however, the microprocessor will load the first byte into the lower eight bits of the register and the second byte into the higher eight bits of the register. In this case, the data must be arranged in two-byte groups ordered as the least significant byte and the most significant byte, as shown in Fig. 7–3.

MULTIPLE-PRECISION MULTIPLICATION

Multiplication using many bytes is very hard to implement. One of the reasons for this is that most microprocessors do not have "wide enough" registers to work with for such multiplications. About the best that can be done is to multiply a two-byte multiplicand by a two-byte multiplier to yield a four-byte product.

Another method of achieving multiple-precision multiplication is to take advantage of the **expansion** of $(A + B) \times (C + D)$. The expansion of $(A + B) \times (C + D)$ is $A \times C + B \times C + B \times D + A \times D$.

Suppose that we want to multiply the two unsigned numbers of 778 and 1066. Both of these can be held in sixteen bits each, as we know. The numbers appear as shown in Fig. 7–4. Note an interesting thing. Each number can be split into two parts—a high-order eight bits, and a low-order eight bits. The high-order eight bits of 778, for example, equal 3×256, and the low-order eight bits equal 10. The high-order eight bits of 1066 equal 4×256, and the low-order eight bits equal 42. We could express 778×1066 as:

$$778 \times 1066 = ([3 \times 256] + 10) \times ([4 \times 256] + 42)$$

The expansion of this is:

$$([3 \times 256] \times [4 \times 256]) + (10 \times [4 \times 256]) + ([3 \times 256] \times 42) + (10 \times 42)$$

```
           3  *  256   +    10      =   778
         ┌──────┴─────┐ ┌────┴─────┐
          00000011      00001010


          00000100      00101010
         └──────┬─────┘ └────┬─────┘
           4  *  256   +    42      =  1066
```

Fig. 7–4. Multiple-precision expansion.

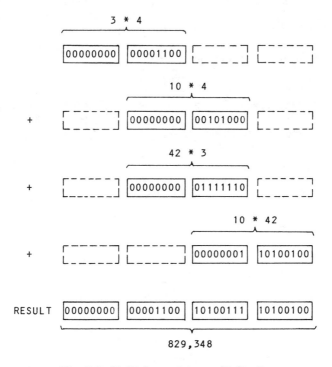

Fig. 7–5. Multiple-precision multiplication.

The first term, $3 \times 256 \times 4 \times 256$, is the same as 3×4 shifted left sixteen bits. The second term is the same as 10×4 shifted left eight bits. The third term is the same as 42×3 shifted left eight bits. The fourth term is a simple multiplication of 10×42. In fact, to compute the product, all we must do is:

1. Clear a 32-bit (four-byte) product.
2. Multiply 3×4 and add the result to the first two bytes of the product.
3. Multiply 10×4 and add the result to the second and third bytes of the product.
4. Multiply 42×3 and add the result to the second and third bytes of the product.
5. Multiply 10×42 and add the result to the third and fourth bytes of the product.

This procedure is shown in Fig. 7–5. It works, of course, not only for this example but for any 16-bit multiplier and multiplicand, and also for larger values. Split the operands into as many parts as necessary, perform four multiplications, align the results and add, and you will have the product.

Unfortunately, multiple-precision divisions cannot be as easily factored. Here again, there are usually not enough registers in the microprocessor and they are not "wide" enough to handle the divide of many bytes effectively. We'll look at division of larger numbers, with some loss of precision, by the floating-point method in Chapter 10. In the meantime, try the following exercises to reinforce what you have learned in this chapter.

EXERCISES

1. What is the largest value that can be held in 24 bits?

2. Add the four-byte multiple-precision operands given below:

```
  00010000  11111111  01011011  00111111
+ 00010001  00000000  11111111  00000001
```

3. Subtract the four-byte multiple-precision operands given below:

```
  00010000  11111111  01011011  00000000
- 00010000  00000000  10000001  00000001
```

4. Take the two's complement of the following four-byte multiple-precision operand:

```
00111111  10101000  10111011  01111111
```

5. Perform a logical shift right of the following four-byte multiple-precision operand:

```
00111111  10101011  10111011  01111111
```

CHAPTER **8**

Fractions and Scaling

We've done a lot of talking about binary numbers in the course of this book, but all of the numbers have been **integer** values. In this chapter, we'll see how fractions can be represented in binary notation, and how numbers can be scaled to represent mixed numbers.

BIG ED WEIGHS THE NUMBERS

"Are you the owner?" asked a chunky individual in a green tie, red and white checked jacket, beige pants, and scuffed black shoes.

"Yes, I am," said Big Ed, shaking his head over the lack of a handkerchief in the vest pocket of the stranger's coat. Certainly not dressed for success, he thought to himself.

"Well, hi. I'm John Upchuck of Acme Sales. I have a sample of our new Binary Scale Mark II here which is specifically geared to restaurants like your own. This little gadget is not a slicer, not a dicer, not a cutter . . . er . . . I'm sorry, I got off on the wrong pitch. What I have here is a scale that you can use to accurately weigh your patron's meat portions. I see that you advertise "twelve ounces of utility beef" for your Big Edburger. Well, this little gadget will make certain that every Big Edburger is exactly twelve ounces—no more, no less."

"How does it work?" asked Ed intrigued by the catchword "binary."

"Let me show you. You see, there are eight weights. The first weight is 8 ounces. The next is 4 ounces. The next is 2 ounces. The next is 1 ounce." (See Fig. 8–1.)

"That seems vaguely reminiscent of something . . . ," mused Ed.

"Well, continuing, the next weight is ½ ounce, the next weight is ¼ ounce, the next ⅛ ounce, and the last weight is 1⁄16 ounce. A total of 8

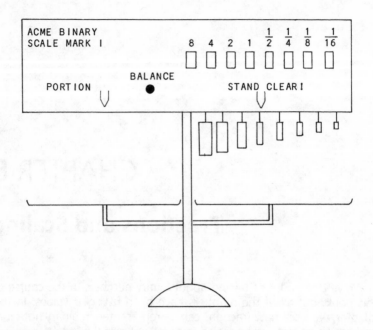

Fig. 8–1. The Binary Scale.

weights, allowing you to weigh any portion of meat or vegetable between $\frac{1}{16}$ ounce and 15 and $^{15}/_{16}$ ounces."

"Now, the operation is simple. You put the meat or portion into the pan on the left side of the scale. You now press one or more of the buttons on the front. You can see that they are labeled 8, 4, 2, 1, ½, ¼, ⅛, and ⅟16. Each time a button is pushed, the weight corresponding to the button is dropped onto the right-hand scale pan and a light illuminates the button. Push the same button again, and the weight is removed from the pan and the light goes out. Here, suppose that you want to weigh out 12 and ¼ ounces of your meatloaf. Put the meatloaf here . . . and, now, press the 8, 4, and ¼ buttons."

As he did so, three weights labeled 8, 4, and ¼ dropped onto the pan, and the lights above the three buttons illuminated. The salesman trimmed the meat until a light marked "BALANCE" came on in the middle of the panel.

"Well, that's very nice," said Ed. "By the way, I've been meaning to ask—what was the Mark I model like?"

"The Mark I model was an early design. It was calibrated in units of ⅟16 of an ounce. The panel was marked like this. . . ." He grabbed a piece of nearby tablecloth, and started drawing furiously (Fig. 8–2).

"The front panel, as you can see, was labeled 128, 64, 32, 16, 8, 4, 2, 1, representing $^{128}/_{16}$ths of an ounce, $^{64}/_{16}$ths ounce, $^{32}/_{16}$ths ounce,

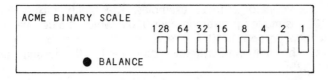

Fig. 8–2. The earlier model.

¹⁶⁄₁₆ths ounce, ⁸⁄₁₆ths ounce, ⁴⁄₁₆ths ounce, and ¹⁄₁₆th ounce. However, it proved to be too complicated for nontechnical people to use, as they had to multiply the number of ounces required by 16 and, then, had to punch the button combinations. People that used the machine started referring to this process as **scaling up** in kind of a derisive way. The Model II is much easier to use."

"Well, would you like me to leave one and bill you later? . . ."

"No, not right now. Thanks much for the demonstration, though. Here, take along one of my meatloaf sandwiches—it'll match your green tie."

FRACTIONS IN BINARY

Binary fractions have a format similar to decimal fractions. There is a binary point instead of a decimal point, which separates the integer portion of the binary number from the fractional part (Fig. 8–3). The bit position immediately to the right of the binary point represents a weight of ½, the next bit position is ¼, the next bit position is ⅛, the next ¹⁄₁₆, and so forth. Whereas the integer bit positions are powers of two, the fractional bit positions are **reciprocals** of powers of two— ½ ↑ 1, ½ ↑ 2, ½ ↑ 3, and so forth.

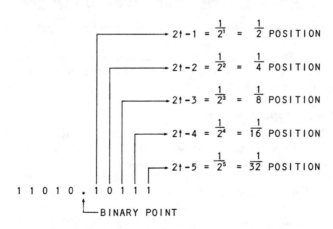

Fig. 8–3. Binary mixed number representation.

Let's consider some more examples. The binary mixed number 0101.1010 is made up of the integer 0101, which is a decimal 5, and the fractional portion .1010. This represents ½ + ⅛, or ⁴⁄₈ + ⅛, which equals ⅝. The total number, then, is equivalent to the decimal number 5⅝ or 5.625.

The binary mixed number 0111.0101011 is made up of the integer 0111, or decimal 7, and the fractional part .0101011. The fractional part is ¼ + ¹⁄₁₆ + ¹⁄₆₄ + ¹⁄₁₂₈, or ³²⁄₁₂₈ + ⁸⁄₁₂₈ + ²⁄₁₂₈ + ¹⁄₁₂₈ = ⁴³⁄₁₂₈. The total mixed number is, therefore, 7.3359375.

To convert from a fractional binary number to decimal, convert the integer portion by any of the methods discussed previously. Next, convert the fractional portion as if it were an integer number. For example, convert 0101011 to 43 as if it were an integer. Now, count the number of bit positions in the fraction and raise the number 2 to that power. Here, the number of bit positions is 7 and 2 raised to the seventh power, or 2 ↑ 7, is 128. Divide 128 into 43. Dividing 43 by 128 gives us ⁴³⁄₁₂₈, the same value that we obtained by adding the separate fractions. This process is shown in Fig. 8–4.

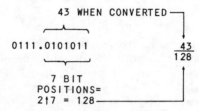

Fig. 8–4. Conversion of a binary fraction.

WORKING WITH FRACTIONS IN BINARY

There are several different ways to process mixed numbers containing fractions in microcomputers. BASIC, of course, uses **floating-point** number routines, which automatically process mixed numbers, but we're primarily talking about a machine-language level here, or perhaps a specialized BASIC code.

Keeping a Separate Fraction

The first way to handle fractions is to keep the fractional part and integer part of a mixed number separate. This scheme is shown in Fig. 8–5, where two bytes in memory hold the integer portion, and an additional one byte holds the fractional part. The binary point is permanently fixed between the integer and fractional part of the number. The integer part and fractional part are processed separately.

The maximum positive number that can be held in this scheme is +32,767.9960 . . . , represented by 01111111 11111111 in the

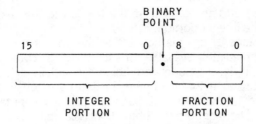

Fig. 8–5. Handling binary mixed numbers.

integer portion and 11111111 in the fractional part. The dots behind the number indicate that the number has additional digits. The maximum negative number that can be represented is −32,768, and is represented by 10000000   00000000 in the integer portion and 00000000 in the fractional portion. To find the magnitude of a negative number in this form, use the two's complement rule discussed earlier. Change all the zeroes to ones, all the ones to zeroes, and add one. In this case, **add one to the least significant bit of the**

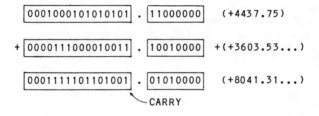

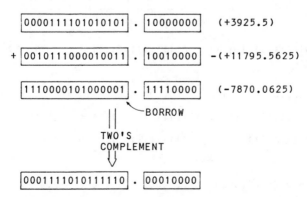

Fig. 8–6. Addition and subtraction of mixed numbers.

fraction and add any carry from the fraction to the next higher bit position of the integer portion.

The advantage of this method is that the fractional portion of the number is immediately available for such things as **rounding** to the nearest cent, if dollar amounts are being handled. In fact, the number can be treated as a 24-bit number in two parts when it comes to additions and subtractions. The fractional part is computed first, and any carry (addition) or borrow (subtraction) is carried over to the integer portion. Fig. 8–6 shows an example of an addition and subtraction using this approach.

Scaling

Another scheme uses a less obvious fixed point. This scheme is similar to the one used by the Acme Binary Scale Mark I. Each number is **scaled up** by a certain amount. The binary point, though not as well defined as when it was on a byte boundary, is there in the programmer's mind.

Here's an example. Suppose that we are using 16-bit numbers. We decide that we want to have four fractional bits. This decision is based upon our requirements for **resolution.** We know that four fractional bits will enable us to get within $\frac{1}{16}$ of the actual value, and this is adequate for our needs.

The layout of our "scaled up by 16" numbers will appear as shown in Fig. 8–7. There are twelve bits of integer, an invisible binary point between bits 4 and 3, and four bits of fraction. The maximum positive

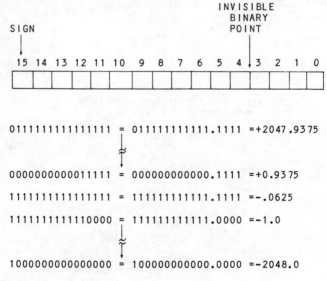

Fig. 8–7. Scaling example.

number that we can handle is 01111111 11111111, which is +2047.9375; the maximum negative number we can handle is −2048.0000, which is 10000000 00000000.

Before we can process any numbers from the outside world, we must scale them up to this fixed-point representation. We do this by entering a valid number in the range of −2048.0000 to −2047.9375 and, then, multiplying by 16. Entering 1000.55, for example, results in 16008.8. We truncate the .8 (drop it), and enter the number 16008 as

ORIGINAL NUMBERS

```
5.25    0101.01
2.5     0010.10
```

SCALED BY 4

```
010101  (5.25 X 4 = 21)
001010  (2.5  X 4 = 10)
```

ADDED

```
010101
001010
011111  = 0111.11 = 7.75
```

SUBTRACTED

```
010101
001010
001011  = 0010.11 = 2.75
```

MULTIPLIED

```
   010101
   001010
   0101010
  101010
11010010  = 110100.10 = 52.5!!

          = 1101.0010 = 15.125
```

DIVIDED

```
            10    R1
001010 |010101
       10, = .10 = .5!!

       1000, = 10.00 = 2.0
```

Fig. 8–8. Operations involving scaling.

a 16-bit binary number. (The number reconverted is $^{16008}\!/_{16}$ = 1000.5, thus losing some accuracy, as we expected.) Other numbers are scaled up in the same fashion.

We then go ahead and add, subtract, multiply, and divide those numbers in any operation that we want. Now comes the rub. If we add and subtract numbers scaled in this fashion, the binary point remains fixed at its position with no problem. **However, multiplication and division move the binary point!** Examples are shown in Fig. 8–8. We must keep track of the point for multiplication and division. Each multiplication moves the point of the result two more positions to the left. Each division moves the point two positions to the right.

At the end of the processing, after the point location has been adjusted for multiplication and division, we reconvert the numbers back into the actual values by dividing by 16. Since multiplying by 16 and dividing by 16 can easily be handled using four 1-bit shifts, the conversion and reconversion are easy. Typical processing for numbers in this format is shown in Fig. 8–9.

Although having to keep track of the binary point is tedious, the

```
ENTER AND
MULTIPLY +36.75 BY +20.25


ENTER

    +36.75 ──▶ 00100100.110 X 16 =
               0000001001001100 (+100.75 SCALED UP)

    +20.25 ──▶ 10100.01 X 16 =
               0000000101000100 (+20.25 SCALED UP)


MULTIPLY

                  0000001001001100
                  0000000101000100
                 000000100100110000
                0000001001001100000
               00000010010011000
               0000001011101000 00110000

                        MOVE POINT
                        FOR MULTIPLY


RECONVERT

    0000001011101000.0011
            │
            ▼
    744.1875
```

Fig. 8–9. Processing scaled numbers.

advantage of this method is that we can handle both the fractional and integer portions of mixed numbers at the same time without having to process the fractional portion separately and "propagate the carry" to the integer portion. Use this approach for any processing that requires a fixed set of operations that are few in number.

Conversion of Data

If you have read this chapter carefully, you'll probably have a significant question at this point. How do we convert the data in character format from the outside world into binary form, scaled or not, and then reconvert it back to a form suitable for display on a video display or printer? We'll answer that question in the next chapter. In the meantime, to keep you honest, we've included some self-testing exercises on fractions and scaling.

EXERCISES

1. What are the equivalent decimal fractions of:

    .10111, .1, .1111, .0000000000001

2. Convert the following decimal fractions to binary values:

    .365, .3125, .777

3. What are the equivalent decimal numbers for the following signed binary mixed numbers:

    00101110.1011, 10110111.1111

4. Scale up these numbers by 256 and show the result as 16-bit signed binary numbers:

    100, −99

5. The following signed numbers are scaled up by eight. What decimal mixed numbers are represented?

    01011100, 1010110101101110

CHAPTER **9**

ASCII Conversions

So far we've done a lot of talking about processing binary data in various formats. How, though, does the data get into the computer? Certainly, some of it is stored in the program as **constant data,** but other data must be input from the real world from a keyboard or terminal and, then, output back for display or printing. In this chapter, we'll discuss how data is converted between its real world form and its computer representation.

BIG ED AND THE INVENTOR

Big Ed had just opened the restaurant. He was busy getting ready for the onslaught of hungry engineers, programmers, marketeers, and computer scientists from the many microcomputer and semiconductor manufacturers around his restaurant.

"Am I late?" puffed a small pudgy man with a chalk-stained suit and a small bow tie.

"Oh, hi. Late for what?"

"Late for lunch. I wanted to get here in time for the lunch crowd from the microcomputer companies."

"No, you're just in time. They should be arriving any minute. Are you meeting someone?"

"No, I wanted to try to strike up some friendships with some of the engineers. My name is Anton Slivovitz. I'm an inventor, and I wanted to get some support for my new invention."

"Well, I can introduce you to some of our regulars. What kinds of inventions do you invent?"

"Mainly high-technology stuff. I invented a saser."

"A saser? Is that like a laser?"

"Well, kind of. It's sound amplification by stimulated emission of radiation. It produces coherent sound waves of one frequency. I thought possibly I could develop it into a death ray of some type, but when I tried it out in a crowded shopping center, not a soul was affected. I also invented a binacus."

"Oh, oh . . . uh . . . , what's your latest invention?"

"The one I'm going to try to push today is a uniform code for microcomputer peripheral devices. You see, if all the manufacturers of video display terminals, keyboards, printers, plotters, and other character-oriented devices used the same codes, then you could easily use **any** peripheral device with any other one, or with any microcomputer system. I've developed what I call my Anton Slivovitz Code for Internal Information, or ASCII, for short. It represents all the alphabetic characters (upper and lower case, of course), numeric digits, and special characters, such as the pound sign, the dollar sign, and the percent sign. It even has provision for special codes."

As Anton was describing his code, Bob Borrow, an engineer from Inlog, walked in.

"Did you say you developed the ASCII code? May I shake your hand, Mr. . . . er. . . ."

"Slivovitz."

"Slivovitz. And could you autograph this pocket reference card for me?"

"Certainly, I . . . , but . . . , I don't understand. This card says 'ASCII Codes.' And they're the same as mine!"

"Mr. Slivovitz, these **are** the ASCII codes!"

"You mean that someone else already uses these?"

"Just about everybody but IBM. And you know that old joke about where a big gorilla sleeps. Hee, hee."

"Oh, my goodness. Well, back to the saser. . . ."

So saying, the dejected inventor made a hurried retreat out the door.

ASCII CODES

As you might have surmised from the above, ASCII codes are a standard set of 7-bit codes for character data that are used with **peripheral devices** for computers, including microcomputers. The ASCII codes are shown in Fig. 9–1. The most significant bit for all codes is not used, and is usually set to 0. The ones that we're primarily concerned with here are the codes that represent the numerals 0 through 9, the code for a plus sign, the code for a minus sign, and the code for a period. The codes for 0 through 9 are 30H through 39H, respectively, while the code for a plus sign is 2BH, the code for a minus sign is 2DH, and the code for a period is 2EH.

MOST SIGNIFICANT
HEX DIGIT

	0X	1X	2X	3X	4X	5X	6X	7X
X0	//	//	BLNK	0	@	P	///	p
X1	//	//	!	1	A	Q	a	q
X2	//	//	"	2	B	R	b	r
X3	//	//	#	3	C	S	c	s
X4	//	//	$	4	D	T	d	t
X5	//	//	%	5	E	U	e	u
X6	//	//	&	6	F	V	f	v
X7	//	//	'	7	G	W	g	w
X8	//	//	(	8	H	X	h	x
X9	//	//	)	9	I	Y	i	y
XA	LF		*	:	J	Z	j	z
XB	//	//	+	;	K	//	k	//
XC	//	//	,	<	L	//	l	//
XD	CR		–	=	M		m	
XE	//	//	.	>	N	//	n	//
XF	//	//	/	?	O	//	o	//

LEAST
SIGNIFICANT
HEX
DIGIT

LF=LINE FEED
CR=CARRIAGE RETURN

//// VARIES

Fig. 9–1. ASCII codes.

As data is input or output from the microcomputer, a **character string** of these codes is handled. For example, inputting a string of data from a keyboard might result in the codes shown in Fig. 9–2 being stored in a memory **buffer** that is dedicated to the keyboard input. Outputting data to a line printer might take place from a line of character codes in a **print buffer,** as shown in the same figure. The program has to convert the input character codes to binary before processing can occur. After processing is done, the binary results are converted back to character codes for display or printing.

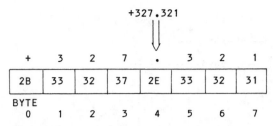

Fig. 9–2. Storage of ASCII input in memory buffer.

CONVERSION FROM ASCII TO BINARY INTEGERS

Let's first consider a conversion from ASCII coding to an integer binary value. Suppose that we have typed a value of five decimal digits from the keyboard. The keyboard driver program has put those characters in a buffer, somewhere in memory.

Before we do the conversion from ASCII characters (representing the digits 0 through 9) into a binary value, we have to define some conditions about the input value. First, we have to define the size of the data that we'll be working with. If we are to be working with 16-bit binary integers, for example, then we'll have to restrict the input value from −32,768 to +32,767. Anything larger would be an invalid input. We'll also have to restrict the input to integer values with no fractions. Finally, we'll have to accept only the characters 0 through 9, and a prefix of "+" or "−".

The conversion would go something like this:

1. Clear a partial product.
2. Multiply the partial product by ten. On the first pass, this would produce zero.
3. Starting from the most significant ASCII character, fetch it and subtract 30H from it.
4. Add the result to the partial product.
5. If this was not the last ASCII character, go to step 2.
6. If there was a prefix of "+", the partial product now holds the 16-bit binary value. If there was a prefix of "−", **negate** the 16-bit partial product to get the negative value.

Of course, there's a lot left unsaid about setting up pointers to the input buffer, getting the character, checking it for a valid value between 30H and 39H, bypassing any sign prefix, and so forth, but this is the general conversion algorithm. A sample of this conversion is shown in Fig. 9–3.

This conversion scheme can be used for any integer input value, although larger values will pose some problems in the holding of the entire partial product in the registers at one time.

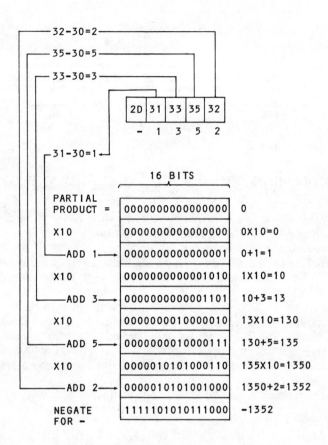

Fig. 9-3. ASCII to binary integer conversion.

CONVERSION FROM ASCII TO BINARY FRACTIONS

When the input string represents a mixed number with an integer, decimal point, and a fraction, the conversion problem is more difficult. Let's take the example shown in Fig. 9-4. The integer portion from the first character to the decimal point is converted as in the preceding scheme, and the result is saved.

The fraction portion is now converted as an integer value (from the first character after the decimal point to the last character). Now, determine how many bits the fraction is to occupy. Scale up by that amount. For example, if the fraction is to occupy eight bits, multiply the result of the conversion by 256; this can be done by simply shifting, or better yet, by adding a byte of zeroes to the result.

Now take the scaled result and perform a division of ten times the number of decimal places in the fraction. If, as in this case, there are four places, divide the scaled-up fraction by 10000. The quotient of the division is the binary fraction to be used, and the binary point is aligned after the original conversion value as shown in Fig. 9–4. If the input value is a negative value, negate both the integer portion and the fraction.

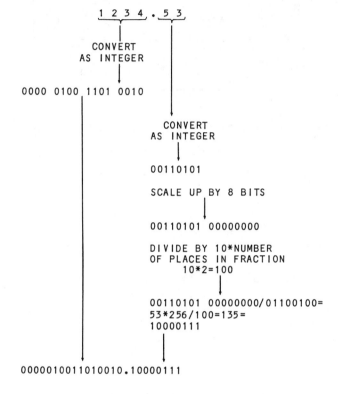

Fig. 9–4. ASCII to binary fraction conversion.

This scheme can be used to convert a mixed number either to the integer/fraction format or to the "implied binary point" format as described in the previous chapter.

CONVERSION FROM BINARY INTEGERS TO ASCII

After processing has been done in the program, the result must be converted to an ASCII string of decimal digits, including a possible sign and decimal point.

Let's first consider conversion of an integer value of eight, sixteen, or some other number of bits. The general approach is to use the "divide

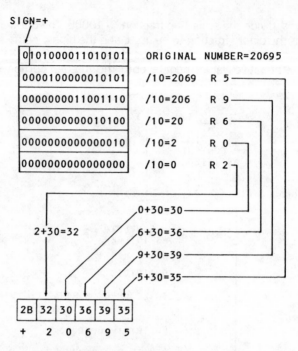

Fig. 9–5. Binary to ASCII integer conversion.

by ten and save remainders" algorithm to get a string of values. Each value can be from zero to nine. As the **last** value represents the first character to be printed, the entire conversion must be done before the translation to ASCII. A sample conversion is shown in Fig. 9–5.

The integer value to be converted is in a 16-bit register in the microprocessor. A test is first made of the sign. If the sign is negative, a **negate** is done to take the absolute value. The sign of the result is saved.

Next, a successive division by ten is done, with the remainders going to a **print buffer** in reverse order. When the quotient is zero, the division is completed and all of the remainders are in the buffer.

Now, each remainder is changed to ASCII by adding 30H to it. The result is an ASCII digit from 0 to 9. After the last digit has been converted, a minus or plus sign is stored at the beginning of the print buffer, depending upon the initial value. The string of ASCII characters can be printed by calling a software print driver.

CONVERSION FROM BINARY FRACTIONS TO ASCII

The number to be output is first tested for sign. If the sign is negative, a flag is set, and the number is **negated** to its absolute value for conversion.

The number is then separated into an integer and fractional part. The number may already be divided in this fashion if a separate integer and fraction are maintained. If the number has been scaled and has an implied fixed binary point, shifting will separate the fraction and integer portion into two component parts.

The integer portion may now be converted by the "divide by ten, save remainders" scheme as shown in Fig. 9–6. The remainders are put in reverse order into a print buffer. The ASCII code for a decimal point is now stored in the print buffer after the last remainder. Now convert each remainder to ASCII by adding 30H.

Finally, the fractional part can be converted. The number of bits in the fraction is found. This is really the number of bits that was maintained while processing of the binary mixed number was taking place. Align these bits in a multiplicand. Now determine the number of decimal places to be output. For each decimal place to be output, multiply

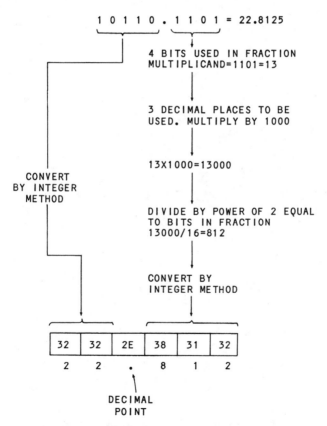

Fig. 9–6. Binary to ASCII fraction conversion.

by a power of ten. In this case, three decimal places are to be used, so the fractional portion is multiplied by 1000.

Next, divide the result of the multiplication by the power of two equal to the number of bits used in the fraction. In this case, four bits are used, so sixteen would be divided into the result of the multiplication. This division may be done by shifting. Discard the bits shifted out. The resulting number can now be converted to ASCII by the method used for integers that was described above—"divide by ten and save remainders" in reverse order and, then, add 30H to convert to ASCII. Add a negative sign in ASCII at the beginning of the print buffer and use a print driver to display the final result.

If you are wondering why floating point is used, the preceding conversion algorithms might shed some light upon why a generic way to handle a wide range of numbers in standard format might be required. Floating-point operations are described in the next chapter. They are tedious, but less so than a program that performs computations for a variety of scaled numbers like the preceding program!

The following exercises will help in your appreciation of the next chapter.

EXERCISES

1. A buffer holds the ASCII values shown below. What decimal numbers are represented?

> 2B, 30, 31, 32, 33, 39, 31
> 2D, 30, 39, 38
> 2D, 30, 31, 2E, 35, 32
> 2B, 2E, 37, 36, 38

2. Convert the following numbers into ASCII representation (with sign):

> +958.2, −1011.59

CHAPTER **10**

Floating-Point Numbers

Floating-point numbers in microcomputers express numbers in a form of **scientific notation,** where each number is made up of a **mantissa** and a power of two. In this chapter, we'll see how floating-point operations work.

. . . AND 3000 COMBINATION PLATES FOR THE MOTHER SHIP . . .

It was early evening, and Big Ed was just getting ready to close down the restaurant. Half a dozen computer engineers were whisked out by Ed as he locked the door to the building and escorted them out into the clear California night.

"Well, g'night boys, see. . . . What the heck is that!"

All heads turned in the direction Ed was pointing. A large object with pulsating orange lights was hanging almost directly overhead Big Ed's. A faint humming sound could be heard.

"It's a UFO! I've finally seen one!" said one of the engineers, excitedly.

Eyes bulged as the object hung there.

"Don't just stand there! Let's try to photograph it, or communicate with it, or **something**!"

"I don't know—remember what happened in *War of the Worlds*!"

"I've got a flashlight in the car. Hang on, I'll get it," said one of the more adventurous engineers.

He ran to his car and came back with a long powerful flashlight.

"You'd better hope they don't think it's some kind of laser, Frank," warned one.

"Listen, we've got to try this. I've been following Carl Sagan. I know what to do. I'll send it a number and see if it responds!"

He pointed the flashlight directly at the object and flicked the button on and off.

"Nine flashes—let's see if it comes back with the same number!"

To everyone's excitement, a bright, directional light beam shot back a long flash, a short flash, a short flash, a long flash, and stopped.

"That's not nine!" said one of the engineers.

"Yes it is—it's nine in binary. They're testing us to see if we've progressed beyond punched-card equipment! Let's try something else. Let's try a Fibonacci series. . . ."

He sent a "1, 1, 2, 3, 5, 8, 13, 21, 34" and then waited. Almost immediately, a "long, long, short, long, long, long" came back.

"That's 110111, for the next term of 55 in the series!" shouted the engineer with the flashlight.

"How about pi," suggested one of the group.

"Yeah, let's try that. Three flashes, one flash, four flashes, one flash . . . , there, that's the first four digits. . . ."

All watched as the object responded with a series of flashes.

"Take these down, Roy," said the engineer with the flashlight.

The UFO sent a burst of flashes, and then the sequence stopped.

"Boy, this doesn't make any sense at all!" said the engineer who was recording the response. What he had recorded is shown in Fig. 10–1.

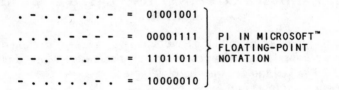

Fig. 10–1. The UFO's response.

"Wait! I've got it! This is pi in floating-point notation!"

The engineers grouped around the message excitedly. In the course of the next ten minutes, several other series were sent and received.

"What's it sending now?"

"I don't know, it's different from the rest. Hey, wait! This is ASCII. It reads . . . , uh . . . , 'CAN YOU READ THIS?' Signal back 'YES'."

The engineer with the flashlight answered with a "YES" in ASCII.

"Now what's it sending?" asked one of the engineers, as the other converted the incoming message to text.

"It looks like a take-out order," cried Ed, glancing over the shoulder

---

*Microsoft is a trademark of Microsoft, Inc.

of the engineer who was decoding the text. " 'FOUR REUBEN SAND-WICHES, TWO ORDERS FRIES, FOUR COKES.' No wonder they stopped here. Well, a customer is a customer. . . ."

Ed turned back to the restaurant, and started unlocking the door. . . .

SCIENTIFIC NOTATION VS. FLOATING POINT

Floating point is really just a binary version of **scientific notation.** Scientific notation can express very large and very small numbers easily, in a uniform format that makes computations simpler. In scientific notation, a number is represented by a mixed number and a power of ten. The number is adjusted so that it has a one-digit integer portion and a fraction.

Take the example of the number of square inches in a square mile. Admittedly, this isn't something that you would normally be too concerned with, unless you're subdividing tracts for ant colonies, but it shows how easy scientific notation is to work with. There are 12 inches in a foot. Within one square foot, therefore, there are 12 × 12 or 144 square inches. Converting to scientific notation goes like this: 144 × 10 ↑ 0 = 144, as any number to the 0 power is one. **Normalizing** the mixed number portion, we move the decimal point two digits over, so that it is between the 1 and 4. For every shift to the left we add one to the **exponent,** or power of ten, so we have:

$$144 = 144 \times 10 \uparrow 0 = 14.4 \times 10 \uparrow 1 = 1.44 \times 10 \uparrow 2$$

The last figure is the normalized, or standard, form for scientific notation. Now, we want to find the number of square feet in a square mile. We know there are 5280 feet per mile, so there must be 5280 × 5280 square feet in a mile. Let's convert 5280 to scientific notation before we proceed.

$$5280 = 5280 \times 10 \uparrow 0 = 528.0 \times 10 \uparrow 1 =$$
$$52.80 \times 10 \uparrow 2 = 5.280 \times 10 \uparrow 3$$

To find the final answer, the number of square inches in a square mile, we can say:

number of sq. inches/sq. mile =
$$5.280 \times 10 \uparrow 3 \times 5.280 \times 10 \uparrow 3 \times 1.44 \times 10 \uparrow 2$$

Now for the profound part. The powers of identical bases are added for multiplication of numbers, and subtracted for division. Since all the numbers use a power of 10, we can simply add the 3 from 10 ↑ 3, the 3 from 10 ↑ 3, and the 2 from 10 ↑ 2. We now have

number of sq. inches/sq. mile =
$$5.280 \times 5.280 \times 1.44 \times 10 \uparrow 8$$

which turns out to be 40.14 × 10 ↑ 8. To express this as a number without an exponent, shift the decimal point to the right, and subtract one from the exponent as you do so. Add zeroes if necessary.

number of sq. inches/sq. mile =
40.14 × 10 ↑ 8 = 401.4 × 10 ↑ 7 =
4014 × 10 ↑ 6 = 40140 × 10 ↑ 5 =
401400 × 10 ↑ 4 = 4014000 × 10 ↑ 3 =
40140000 × 10 ↑ 2 = 401400000 × 10 ↑ 1 =
4014000000 × 10 ↑ 0 = 4,014,000,000

Exponents may also be used in scientific notation to express very small numbers. Let's calculate how many angels can dance on the head of a pin. We'll use the standard angelic dimensions of one angel per $\frac{1}{11,000}$ square inch, or 0.0000909. The size of an ordinary common pin head (according to the American Association of Common Pin Manufacturers) is 0.000965 square inch. Expressing both numbers in scientific notation, we have:

area for one angel = .0000909 =
.0000909 × 10 ↑ 0 = 0.000909 × 10 ↑ −1 =
0.00909 × 10 ↑ −2 = 0.0909 × 10 ↑ −3 =
0.909 × 10 ↑ −4 = 9.09 × 10 ↑ −5

and

area for head of pin = .000965 =
.000965 × 10 ↑ 0 = 0.00965 × 10 ↑ −1 =
0.0965 × 10 ↑ −2 = 0.965 × 10 ↑ −3 =
9.65 × 10 ↑ −4

In the above conversions, we subtracted one from the exponent every time we shifted the decimal point to the right. The negative power of 10 lets us represent reciprocal powers of ten—$\frac{1}{10}$, $\frac{1}{100}$, $\frac{1}{1000}$, and so forth.

Now to find the number of angels doing the microcomputer waltz on the head of a pin, we divide the size of the pin by the size of an angel:

number of angels on the head =
(9.65 × 10 ↑ −4)/(9.09 × 10 ↑ −5)

Here we apply the rule that if the bases are identical, then we can subtract the exponents for the division:

number of angels on the head =
9.65/9.09 × 10 ↑ (−4 + 5) =
9.65/9.09 × 10 ↑ 1 = 1.06 × 10 ↑ 1 =
10.6 × 10 ↑ 0 = 10.6 angels

Using standard scientific notation has made the calculations much easier, as all of the values have been in standard base-10 format, and we can add exponents when multiplying and subtract them when dividing.

It turns out that adding and subtracting two numbers in scientific notation is more complicated in some respects than multiplying them. Consider the two numbers $3/32$ and $6/60$. In scientific notation, these are

$$3/32 = 0.09375 = 0.09375 \times 10 \uparrow 0 =$$
$$.9375 \times 10 \uparrow -1 = 9.375 \times 10 \uparrow -2$$

and

$$6/60 = 0.1 = 0.1 \times 10 \uparrow 0 = 1.0 \times 10 \uparrow -1$$

Now, in order to add or subtract two numbers in scientific notation, **their exponents must be the same.** One or the other of the numbers, therefore, must be adjusted to make both exponents equal. We can do this either by moving the decimal point of $9.375 \times 10 \uparrow -2$ to the left and adding one to make $.9375 \times 10 \uparrow -1$, or by moving the decimal point of $1.0 \times 10 \uparrow -1$ to the right and subtracting one to make $10.0 \times 10 \uparrow -2$. Once the exponents are equal, we can add or subtract.

$$
\begin{array}{rl}
3/32 + 6/60 = & 9.375 \times 10 \uparrow -2 \\
+ & 10.0 \quad \times 10 \uparrow -2 \\
\hline
& 19.375 \times 10 \uparrow -2 = \\
& 1.9375 \times 10 \uparrow -1 = \\
& 0.19375 \times 10 \uparrow 0 = 0.19375
\end{array}
$$

Right about now, you're probably asking yourself, "Why didn't he just add the messy things in 'regular' notation?" With many computations, it makes more sense to convert to scientific notation; it's easier to keep track of the decimal point, for one thing.

USING POWERS OF TWO IN PLACE OF POWERS OF TEN

It turns out that the rules for adding, subtracting, multiplying, and dividing numbers in the same base work for **any** base. We could have just as easily used base-8, base-11, or (he said, triumphantly) base-2!

Let's define a standard base-2 format for numbers. This will be the same format that is used in many microcomputers. To begin with we have to choose a range of numbers. If we use base-2 format (some larger machines use base 16), we will then need a place to keep an exponent. We can allocate some convenient number of bits for the

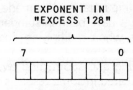

EXPONENT IN
"EXCESS 128"

7　　　　　　　0

VALUE	AFTER ADJUSTMENT BY −128 SUBTRACTION	POWER OF 2
11111111	01111111	+127
11010000	01010000	+80
10000101	00000101	+5
10000000	00000000	0
01111111	11111111	−1
01111110	11111110	−2
00001111	10001111	−112
00000000	10000000	−128

Fig. 10–2. Exponent format and values.

exponent. Since everything in microcomputers seems to be built around bytes, let's use one byte for the exponent. This will give us eight bits, allowing us a range from $2 \uparrow 0$ to $2 \uparrow 8 - 1$, or from 0 to $2 \uparrow 255$. This would allow us to represent numbers that would be equal to about $5.79 \times 10 \uparrow 76$.

One moment, though. We need negative powers of two, also, as we need to represent small values. We'll make the eight bits of an exponent an **excess 128 code.** This means that we'll add 128 to the

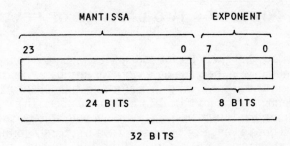

Fig. 10–3. Floating-point format, first try.

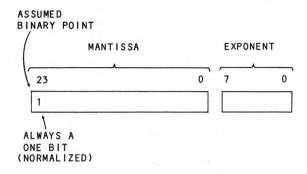

Fig. 10-4. Floating-point format, second try.

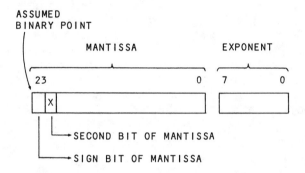

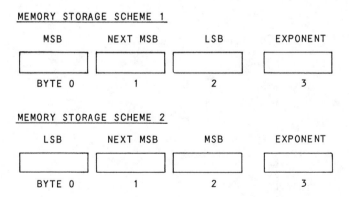

Fig. 10-5. Floating-point format, final version.

exponent value to obtain the number stored in the exponent byte, and subtract it from any result to get the true exponent. The excess 128 code is done to simplify handling the floating-point number as a single entity. Fig. 10-2 shows the exponent format and some sample values.

We now have a range of exponents from $2 \uparrow -128$ to $2 \uparrow +127$ (3.4 $\times 10 \uparrow -38$ to $1.7 \times 10 \uparrow 38$).

Now for the **mantissa,** the number that is multiplied by the power of two. Rather than defining a range, the mantissa defines a **precision.** We know that two bytes, or sixteen bits, gives us values from 0 to 65,535 and about 4½ decimal digits. This doesn't seem quite large enough for many problems. To keep things in multiples of bytes, we'll have to go to three bytes, or 24 bits. That will give us from 0 to 16,777,215, or about 7 decimal digits of precision, which is probably a good compromise between storage requirements and precision.

What we now have as an approximate format is shown in Fig. 10–3—three bytes of mantissa, and one byte for the exponent. What can we say about **normalization**? The guiding rule here is that we would like to retain as much precision as possible. We do that by keeping as many significant bits as possible. Since we have a limited number of bits to hold the mantissa, we must get rid of all extraneous bits and just keep the significant ones. We do that by normalizing the mantissa so that the first "1" bit is immediately to the right of an assumed binary point just in front of the first mantissa bit, as shown in Fig. 10–4. That way, we are assured of maximum precision as there are no extraneous zeroes in front of the first 1 bit and no **truncated** portion of the mantissa at the end. The first bit in the mantissa, therefore, is always a 1.

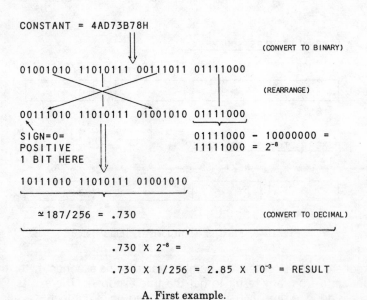

A. First example.

Fig. 10–6. Floating-point

Let's see. We now have one byte for the exponent and a three-byte normalized mantissa. What have we forgotten? It appears that we can express positive numbers, but no signed numbers. The sign bit has disappeared with normalization. We definitely need a sign bit, unless we choose to limit our BASIC users to positive numbers only. . . .

One solution goes like this: As every floating-point number is going

CONSTANT = 00000081H

(CONVERT TO BINARY)

00000000 00000000 00000000 10000001

(REARRANGE)

00000000 00000000 00000000 10000001

SIGN=0=
POSITIVE
1 BIT HERE

10000001 – 10000000 =
00000001 = 2^1

10000000 00000000 00000000

1/2

(CONVERT TO DECIMAL)

1/2 X 2^1 = 1 = RESULT

B. Second example.

CONSTANT = 64269987H

(CONVERT TO BINARY)

01100100 00100110 10011001 10000111

(REARRANGE)

10011001 00100110 01100100 10000111

SIGN=1=
NEGATIVE
1 BIT HERE

10000111 – 10000000 =
00000111 = 2^7

10011001 00100110 01100100

$\simeq$ 153/256 = .598

(CONVERT TO DECIMAL)

.598 X 128 = -76.54 = RESULT

ADD SIGN

C. Third example.

constants in memory.

to be normalized anyway, why not let the first mantissa bit represent the sign bit? The remainder of the number will be held as a two's complement number. We'll just throw away the first 1 bit as we know what it is anyway. The final format for floating-point calculation now has the form shown in Fig. 10–5.

You'll note in Fig. 10–5 that there are two schemes for storing floating-point numbers. This is due to the way different microprocessors store 16-bit values. If data is stored in memory by two 16-bit stores (Z-80 microprocessor), then the format has the least significant byte first, followed by the most significant byte. This means that the floating-point number is stored as follows: least significant byte of the mantissa, next most significant byte of the mantissa, most significant byte of the mantissa, and exponent byte.

Fig. 10–6 shows three examples of constants stored in memory. Study these to better learn how to decode floating-point formats.

DOUBLE-PRECISION FLOATING-POINT NUMBERS

The range of exponents is probably more than adequate for most processing. Few quantities are greater than 1 to the 39th power or smaller than one part in 1 to the 39th power. The number of digits of precision, however, might be increased if the cost in storage is not too high. Another decimal digit is added for approximately every 3½ bits (3 bits scaled up by 8 and 4 bits scaled up by 16). Therefore, for each two bytes added to the mantissa, about 5 decimal digits are added.

A double-precision floating-point format found in some versions of BASIC adds four more bytes to the mantissa, to make the total precision about 17 decimal digits while retaining the same range of numbers. This scheme is shown in Fig. 10–7.

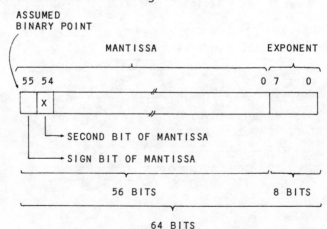

Fig. 10–7. Double-precision floating-point format.

CALCULATIONS USING BINARY FLOATING-POINT NUMBERS

Operations using binary floating-point numbers emulate the scientific notation operations. Two normalized floating-point numbers may be multiplied or divided by performing an integer multiplication on the mantissa, and adding or subtracting the exponents.

Additions or subtractions of two floating-point numbers have to be performed with their exponents adjusted to equal values. As the format of the mantissa allows only fractions less than one, the number with the smaller exponent is adjusted by shifting the mantissa right and adding one to the exponent value for each shift until the exponents are equal.

The algorithms for floating-point operations are rather complicated, and the actual code for floating-point operations is a sizable chunk in a BASIC interpreter. Although we can't go into detail here, we hope that the material in the past two chapters has provided some insight into the general approach to handling fractions, mixed numbers, and floating-point operations.

EXERCISES

1. Convert the following numbers to scientific notation:

    3.141600,   93,000,000,   −186,000,   0.0000135

2. Perform the following operations using scientific notation:

    $$3.14 \times .000152 \times 200,000 = ?$$
    $$3,100,000/.000015 = ?$$

3. Express these powers of two as 8-bit exponent values with excess 128 coding:

    $2 \uparrow 0$, $2 \uparrow 1$, $2 \uparrow -5$, $2 \uparrow 35$, $2 \uparrow -40$, $2 \uparrow 128$, $2 \uparrow -129$

4. Convert these floating-point numbers to decimal values. The mantissa is the most significant byte to the least significant byte, left to right. The exponent is the byte on the extreme right.

    00010000   00000000   00000000   10000111 = ?
    10010011   10000000   00000000   10000111 = ?

APPENDIX A

Answers to Exercises

CHAPTER 1

1. 10100, 10101, 10110, 10111, 11000, 11001, 11010, 11011, 11100, 11101, 11110, 11111, 100000.
2. 53, 16, 85, 240, 14185.
3. 1111, 11010, 110100, 01101001, 11111111, 1110101001100000.
4. 00000101, 00110101, 00010101.
5. 15, 63, 255, 65535. $(2 \uparrow n) - 1$.

CHAPTER 2

1. $9 \times 16 \uparrow 2 + 14 \times 16 \uparrow 1 + 2 \times 16 \uparrow 0$.
2. 0, 1, 2, 3, 4, 5, 6, 7, 8, 9, A, B, C, D, E, F, 10, 11, 12, 13, 14.
3. 5, AH, AAH, 4FH, B63AH.
4. 101011100011, 100110011001, 1111001000110010.
5. 227, 82, 43690.
6. DH, FH, 1CH, 3E8H.
7. FFFFH.
8. 73, 219.
9. 7, 161, 310.
10. It is impossible.
11. 321.

CHAPTER 3

1. $1 + 3 = 4$ (100), $7 + 15 = 22$ (10110), $21 + 42 = 63$ (111111).
2. $2 - 1 = 1$ (1), $7 - 5 = 2$ (10), $12 - 1 = 11$ (1011).
3. Not necessary 111, necessary -86, necessary -128.
4. 11111111, 11111110, 11111101, 11100010, 00000101, 01111111.
5. 00000000011111111 $(+127, +127)$, 1111111110000000 $(-128, -128)$, 1111111110101010 $(-85, -85)$.
6. 1111111011010100 (-300) + 1111111111111011 (-5) = 1111111011001111 (-305).
7. 1111111011010100 (-300) − 1111111111111011 (-5) = 1111111011011001 (-295).

CHAPTER 4

1. $+127+1 = +128$ (10000000, overflow), $+127-1 = +126$ (01111110, no overflow), $-81+(-1) = -82$ (10101110, no overflow).
2. Result = 10000000 (overflow, but no carry), result = 00000000 (no overflow, but carry).
3. Result = 00000000 (Z = 1, S = 0), result = 00110101 (Z = 0, S = 0).

CHAPTER 5

1. 10101111, 11110111.
2. 10100101, 11010111.
3. XXXYYXXX AND 00011000 = 000YY000.
4. 11100001 (-31), 1011 (-5), 01010110 ($+86$).
5. 01011110, 00000001.
6. 10010111, 01000000.
7. C = 0 01011111, C = 1 00000000.
8. 00010111 C = 1, 11000000 C = 0.
9. 00111111 C = 1 (before = $+127$ after = $+63$), 00101101 C = 0 (before = $+90$ after = $+45$), 01000010 C = 1 (before = -123 after = $+66$), 01000000 C = 0 (before = -128 after = $+64$).
10. 11111110 C = 0 (before = $+127$ after = -2), 10110100 C = 0 (before = $+90$ after = -76), 00001010 C = 1 (before = -123 after = $+10$), 00000000 C = 1 (before = -128 after = 0).
11. 00111111 (before = $+127$ after = $+63$), 11000010 (before = -123 after = -62), 11000000 (before = -128 after = -64).

CHAPTER 6

1. 1111111000000001 ($255 \times 255 = 65025$).
2. 1111111111111111, 1111111111111111, division by zero.
3. 1 = negative, 0 = positive by Exclusive-OR of signs of operands.

CHAPTER 7

1. 11111111 11111111 11111111 or 16,777, 215.
2. 00100010 00000000 01011010 01000000.
3. 00000000 11111110 11011001 11111111.
4. 11000000 01010111 01000100 10000001.
5. 00011111 11010101 11011101 10111111 C = 1.

CHAPTER 8

1. .71875, .5, .9375, .00012207 . . .
2. .01011101 . . . , .0101, .1100011 . . .
3. 46.6875, -73.0625.
4. 0110010000000000, 1001110100000000.
5. 11.5, 2642.25.

CHAPTER 9

1. $+012391$, -098, -01.52, $+.768$.
2. 2B 39 35 38 2E 32, 2D 31 30 31 31 2E 35 39.

CHAPTER 10

1. $3.1416 \times 10 \uparrow 0$, $9.3 \times 10 \uparrow 7$, $-1.86 \times 10 \uparrow 5$, $1.35 \times 10 \uparrow -5$.
2. $3.14 \times 1.52 \times 10 \uparrow -4 \times 2.0 \times 10 \uparrow 5 = 9.5 \times 10 \uparrow 1$
   $3.1 \times 10 \uparrow 6 / 1.5 \times 10 \uparrow -5 = 2.0 \times 10 \uparrow 11$.
3. 10000000, 10000001, 01111011, 10100011, 01011000, impossible, impossible.
4. 72.0, -73.75.

APPENDIX B

Glossary of Terms

accumulator—The main register(s) in a microprocessor used for arithmetic, shifting, logical, and other operations. The Z-80 microprocessor has one (A register), while the 6809E microprocessor has two (A and B registers).

add with carry—A machine-language instruction in which one operand is added to another, along with a possible carry from the previous (lower-order) add.

algorithm—A step-by-step process for performing a task.

AND—A bit-by-bit logical operation which produces a 1 in the result bit only if both operand bits are 1s.

arithmetic shift—A type of shift in which an operand is shifted right or left with the sign bit being extended (right shift) or maintained (left shift).

ASCII—A standard code for representation of character data in computers and computer peripheral equipment.

assembly language—A symbolic computer language that is translated by an "assembler program" into machine language, the numeric codes that are equivalent to microprocessor instructions.

base—The "starting point" for representation of a number in written form, where numbers are expressed as multiples of powers of the base value.

binary—Representation of numbers in "base-two," where all numbers are expressed by combinations of the binary digits 0 and 1.

binary digit—The two digits (0 and 1) used in binary notation. Often shortened to "bit."

binary point—The point, analogous to a decimal point, that separates the integer and fractional portions of a binary mixed number.

bit—Contraction of "binary digit."

bit position—The position of a binary digit within a byte or larger group of binary digits. Bit positions in most microcomputers are numbered from right to left, zero through N. This number corresponds to the power of two represented.

borrow—A 1 bit that is subtracted from the next higher bit position.

buffer—A portion of memory dedicated to hold characters (or other data) as they are read in, or used to store characters (or other data) for output.

byte—A collection of eight binary digits. Each memory location in most microcomputers is one byte "wide."

carry—A 1 bit added to the next higher bit position or to the "carry flag."

carry flag—A bit in the microprocessor used to record the carry "off the end" as a result of a machine-language instruction.

clobber—(computerese) To destroy the contents of memory or a register.

Colossus—A British computer used to "crack" German Enigma codes during World War II.

conditional jump—A machine-language instruction that causes a jump to another instruction if a specified flag (or flags) is set or reset.

data—A generic term for numbers, operands, program instructions, flags, or any representation of information using binary ones and zeroes.

displacement—A signed value in machine language used in defining a memory address.

dividend—The number that is divided by the divisor. In A/B, A is the dividend.

divisor—The number that "goes into" the dividend in a divide operation. In A/B, B is the divisor.

double-dabble—A method of converting from binary to decimal representation by doubling a leftmost bit, adding the next bit, and continuing until the rightmost bit has been processed.

eight-by-eight multiply—The multiplication of eight bits by eight bits to produce a 16-bit product.

Enigma—A German cipher machine (World War II).

excess 128 code—A standard way to bias the exponent in Microsoft™ BASIC. The value 128 is added to the actual power of two and then stored as an exponent value in floating-point representation.

Exclusive-OR—A bit-by-bit logical operation that produces a 1 bit in the result only if one or the other (but not both) operand bits is a 1.

---

*Microsoft is the trademark of Microsoft, Inc.

exponent—In this book, usually the power of two of a binary floating-point number.

Fibonacci series—The sequence of numbers 1, 1, 2, 3, 5, 8, 13, 21, 34 . . . , where each term is computed by the addition of the two previous terms.

floating-point number—A standard way of representing a number of any size in computers. Floating-point numbers contain a fractional portion (mantissa) and power of two (exponent) in a form similar to scientific notation.

H—A suffix for hexadecimal numbers.

hex—Abbreviation of hexadecimal.

hexa-dabble—The conversion from hexadecimal to decimal by multiplying each hex digit by sixteen and adding the next hex digit until the last (rightmost) hex digit has been reached.

hexadecimal—The representation of numbers in a "base-sixteen" notation by use of the hexadecimal digits 0, 1, 2, 3, 4, 5, 6, 7, 8, 9, A, B, C, D, E, and F.

Inclusive-OR—A bit-by-bit logical operation which produces a 1-bit result if one or the other operand bits, or both, is a 1.

integer variable—A BASIC variable type. Can hold values of $-32,768$ through $+32,767$ in a two-byte "two's complement" notation.

iteration—One pass through a given set of instructions.

least significant bit—The rightmost bit in a binary value, representing $2 \uparrow 0$.

logical shift—A type of shift in which an operand is shifted right or left, with a zero filling the vacated bit position.

machine language—The string of numeric codes that make up microprocessor instructions. These values are produced by an "assembler program" from assembly language code.

mantissa—The fractional portion of a floating-point number.

minuend—The number from which the subtrahend is subtracted. In $5 - 3 = 2$, 5 is the minuend.

mixed number—A number that consists of an integer and a fraction as, for example, 4.35 or (binary) 1010.1011.

most significant bit—The leftmost bit in a binary value, representing the highest-order power of two. In two's complement notation, this bit is the sign bit.

most significant byte—The highest-order byte. In the multiple-precision number A13EF122H, the two hex digits A and 1 make up the most significant byte.

multiple-precision numbers—Multiple-byte numbers that allow extended precision.

multiplicand—The number to be multiplied by the multiplier. The number "on the top."

multiplicand register—The register used to hold the multiplicand during a machine-language multiplication.

multiplier—The number that is multiplied against the multiplicand. The number "on the bottom."

negation—Changing a negative value to a positive value, or vice versa. Taking the two's complement by changing all the ones to zeroes, all zeroes to ones, and, then, adding a one.

normalization—Converting data to a standard format for processing. In floating-point format, converting a number so that a "significant" bit (or hex digit) is in the first bit (or four bits) of the fraction.

NOT—A logical operation that takes the "inverse" or one's complement.

octal—The representation of numbers using a "base-eight" notation which uses the octal digits 0, 1, 2, 3, 4, 5, 6, and 7.

octal-dabble—The conversion of an octal number to a decimal number by multiplying by eight and adding the next octal digit; continuing until the last (rightmost) digit has been added.

operands—The numeric values used in an add, subtract, or other operation.

OR—See Inclusive-OR.

overflow—A condition that exists when the result of an addition, subtraction, or other arithmetic operation is too large to be held in the number of bits allotted.

overflow flag—A bit in the microprocessor used to record an overflow condition for machine-language operation.

padding—Filling bit positions to the left of a value with zeroes to make a total of eight or sixteen bits.

partial product—The intermediate results of a multiplication. At the end, the partial product becomes the whole product.

partial-product register—The register(s) used to hold the partial results of a machine-language multiplication.

peripheral devices—A generic term for equipment attached to a computer, such as keyboards, disk drives, cassette tape recorders, printers, plotters, speech synthesizers, and so forth.

permutation—Arrangements of things in a definite order. Two binary digits have four permutations—00, 01, 10, and 11.

positional notation—Representation of a number where each digit position represents an increasingly higher power of the base.

precision—The number of significant digits that a variable or number format may contain.

print buffer—A portion of memory dedicated to holding the string of characters to be printed.

product—The result of a multiplication.

propagation—The manner in which a carry or borrow travels to the next higher bit position.

punched-card equipment—Peripheral devices that permit the punching or reading of paper punched cards used to hold character or binary data.

quotient—The result of a divide operation.

register—A fast-access memory location in the microprocessor of a microcomputer. Used for holding intermediate results and, also, used for computation in machine language.

remainder—The amount of dividend remaining after a division has been completed.

residue—The amount of dividend remaining part way through a division.

restoring division—A division in which the divisor is restored if the operation "does not go" for any iteration. A common microcomputer division technique.

rotate—A type of shift in which data is recirculated right or left back into the operand from the opposite end.

rounding—The process of truncating bits to the right of a bit position and adding a zero or a one to the next higher bit position based on the value to the right. Rounding the binary fraction 1011.1011 to two fractional bits, for example, results in 1011.11.

scaling—Multiplying a number by a fixed amount so that a fraction can be processed as an integer value.

scaling up—Refers to a number that has been multiplied by a scale factor for processing.

scientific notation—A standard form for representing any size number by a mantissa and a power of ten.

shift and add—A method in which multiplication is achieved by the shifting and addition of the multiplicand.

sign bit—The leftmost bit (bit position 15 or 7) of a two's complement number. If a zero, the sign of the number is positive. If it is a one, the sign of the number is negative.

sign extension—Extending the sign bit of a two's complement number to the left by duplication.

sign flag—A bit in the microprocessor used to record the sign of the result of a machine-language operation.

sign magnitude—A nonstandard way of representing positive and negative numbers in microcomputers.

signed numbers—Numbers that may be either positive or negative.

significant bits—The number of bits in a binary value after the leading zeroes have been removed.

subtract with carry—A machine-language instruction in which one operand is subtracted from another, along with a possible borrow from the next lower byte.

subtrahend—The number that is subtracted from the minuend. In $5 - 3 = 2$, 3 is the subtrahend.

successive addition—A multiplication method in which the multiplicand is added a number of times equal to the multiplier to find the product.

truncation—The process of dropping bits to the right of a bit position. Truncating the binary fraction 1011.1011 to a number with fraction of two bits, for example, results in 1011.10.

truth table—A table defining the results for several different variables and containing all possible states of the variables.

Turing—An early (1940's) British computer scientist and mathematician.

two's complement—A standard way of representing positive and negative numbers in microcomputers.

unsigned numbers—Numbers that may be only positive. Absolute numbers.

Williams tube—An early type of computer memory based upon the storage of data on the face of a cathode-ray tube; designed by F. C. Williams, Manchester University.

word—A collection of sixteen binary digits. Two bytes.

zero flag—A bit in the microprocessor used to record the zero/nonzero status of the result of a machine-language instruction.

APPENDIX C

Binary, Octal, Decimal, and Hexadecimal Conversions

Binary	Octl	Decl	Hex	Binary	Octl	Decl	Hex	Binary	Octl	Decl	Hex
0000000000	0000	0	000	0000101010	0052	42	02A	0001010100	0124	84	054
0000000001	0001	1	001	0000101011	0053	43	02B	0001010101	0125	85	055
0000000010	0002	2	002	0000101100	0054	44	02C	0001010110	0126	86	056
0000000011	0003	3	003	0000101101	0055	45	02D	0001010111	0127	87	057
0000000100	0004	4	004	0000101110	0056	46	02E	0001011000	0130	88	058
0000000101	0005	5	005	0000101111	0057	47	02F	0001011001	0131	89	059
0000000110	0006	6	006	0000110000	0060	48	030	0001011010	0132	90	05A
0000000111	0007	7	007	0000110001	0061	49	031	0001011011	0133	91	05B
0000001000	0010	8	008	0000110010	0062	50	032	0001011100	0134	92	05C
0000001001	0011	9	009	0000110011	0063	51	033	0001011101	0135	93	05D
0000001010	0012	10	00A	0000110100	0064	52	034	0001011110	0136	94	05E
0000001011	0013	11	00B	0000110101	0065	53	035	0001011111	0137	95	05F
0000001100	0014	12	00C	0000110110	0066	54	036	0001100000	0140	96	060
0000001101	0015	13	00D	0000110111	0067	55	037	0001100001	0141	97	061
0000001110	0016	14	00E	0000111000	0070	56	038	0001100010	0142	98	062
0000001111	0017	15	00F	0000111001	0071	57	039	0001100011	0143	99	063
0000010000	0020	16	010	0000111010	0072	58	03A	0001100100	0144	100	064
0000010001	0021	17	011	0000111011	0073	59	03B	0001100101	0145	101	065
0000010010	0022	18	012	0000111100	0074	60	03C	0001100110	0146	102	066
0000010011	0023	19	013	0000111101	0075	61	03D	0001100111	0147	103	067
0000010100	0024	20	014	0000111110	0076	62	03E	0001101000	0150	104	068
0000010101	0025	21	015	0000111111	0077	63	03F	0001101001	0151	105	069
0000010110	0026	22	016	0001000000	0100	64	040	0001101010	0152	106	06A
0000010111	0027	23	017	0001000001	0101	65	041	0001101011	0153	107	06B
0000011000	0030	24	018	0001000010	0102	66	042	0001101100	0154	108	06C
0000011001	0031	25	019	0001000011	0103	67	043	0001101101	0155	109	06D
0000011010	0032	26	01A	0001000100	0104	68	044	0001101110	0156	110	06E
0000011011	0033	27	01B	0001000101	0105	69	045	0001101111	0157	111	06F
0000011100	0034	28	01C	0001000110	0106	70	046	0001110000	0160	112	070
0000011101	0035	29	01D	0001000111	0107	71	047	0001110001	0161	113	071
0000011110	0036	30	01E	0001001000	0110	72	048	0001110010	0162	114	072
0000011111	0037	31	01F	0001001001	0111	73	049	0001110011	0163	115	073
0000100000	0040	32	020	0001001010	0112	74	04A	0001110100	0164	116	074
0000100001	0041	33	021	0001001011	0113	75	04B	0001110101	0165	117	075
0000100010	0042	34	022	0001001100	0114	76	04C	0001110110	0166	118	076
0000100011	0043	35	023	0001001101	0115	77	04D	0001110111	0167	119	077
0000100100	0044	36	024	0001001110	0116	78	04E	0001111000	0170	120	078
0000100101	0045	37	025	0001001111	0117	79	04F	0001111001	0171	121	079
0000100110	0046	38	026	0001010000	0120	80	050	0001111010	0172	122	07A
0000100111	0047	39	027	0001010001	0121	81	051	0001111011	0173	123	07B
0000101000	0050	40	028	0001010010	0122	82	052	0001111100	0174	124	07C
0000101001	0051	41	029	0001010011	0123	83	053	0001111101	0175	125	07D

Binary	Octl	Decl	Hex	Binary	Octl	Decl	Hex	Binary	Octl	Decl	Hex
0001111110	0176	126	07E	0010111010	0272	186	0BA	0011110110	0366	246	0F6
0001111111	0177	127	07F	0010111011	0273	187	0BB	0011110111	0367	247	0F7
0010000000	0200	128	080	0010111100	0274	188	0BC	0011111000	0370	248	0F8
0010000001	0201	129	081	0010111101	0275	189	0BD	0011111001	0371	249	0F9
0010000010	0202	130	082	0010111110	0276	190	0BE	0011111010	0372	250	0FA
0010000011	0203	131	083	0010111111	0277	191	0BF	0011111011	0373	251	0FB
0010000100	0204	132	084	0011000000	0300	192	0C0	0011111100	0374	252	0FC
0010000101	0205	133	085	0011000001	0301	193	0C1	0011111101	0375	253	0FD
0010000110	0206	134	086	0011000010	0302	194	0C2	0011111110	0376	254	0FE
0010000111	0207	135	087	0011000011	0303	195	0C3	0011111111	0377	255	0FF
0010001000	0210	136	088	0011000100	0304	196	0C4	0100000000	0400	256	100
0010001001	0211	137	089	0011000101	0305	197	0C5	0100000001	0401	257	101
0010001010	0212	138	08A	0011000110	0306	198	0C6	0100000010	0402	258	102
0010001011	0213	139	08B	0011000111	0307	199	0C7	0100000011	0403	259	103
0010001100	0214	140	08C	0011001000	0310	200	0C8	0100000100	0404	260	104
0010001101	0215	141	08D	0011001001	0311	201	0C9	0100000101	0405	261	105
0010001110	0216	142	08E	0011001010	0312	202	0CA	0100000110	0406	262	106
0010001111	0217	143	08F	0011001011	0313	203	0CB	0100000111	0407	263	107
0010010000	0220	144	090	0011001100	0314	204	0CC	0100001000	0410	264	108
0010010001	0221	145	091	0011001101	0315	205	0CD	0100001001	0411	265	109
0010010010	0222	146	092	0011001110	0316	206	0CE	0100001010	0412	266	10A
0010010011	0223	147	093	0011001111	0317	207	0CF	0100001011	0413	267	10B
0010010100	0224	148	094	0011010000	0320	208	0D0	0100001100	0414	268	10C
0010010101	0225	149	095	0011010001	0321	209	0D1	0100001101	0415	269	10D
0010010110	0226	150	096	0011010010	0322	210	0D2	0100001110	0416	270	10E
0010010111	0227	151	097	0011010011	0323	211	0D3	0100001111	0417	271	10F
0010011000	0230	152	098	0011010100	0324	212	0D4	0100010000	0420	272	110
0010011001	0231	153	099	0011010101	0325	213	0D5	0100010001	0421	273	111
0010011010	0232	154	09A	0011010110	0326	214	0D6	0100010010	0422	274	112
0010011011	0233	155	09B	0011010111	0327	215	0D7	0100010011	0423	275	113
0010011100	0234	156	09C	0011011000	0330	216	0D8	0100010100	0424	276	114
0010011101	0235	157	09D	0011011001	0331	217	0D9	0100010101	0425	277	115
0010011110	0236	158	09E	0011011010	0332	218	0DA	0100010110	0426	278	116
0010011111	0237	159	09F	0011011011	0333	219	0DB	0100010111	0427	279	117
0010100000	0240	160	0A0	0011011100	0334	220	0DC	0100011000	0430	280	118
0010100001	0241	161	0A1	0011011101	0335	221	0DD	0100011001	0431	281	119
0010100010	0242	162	0A2	0011011110	0336	222	0DE	0100011010	0432	282	11A
0010100011	0243	163	0A3	0011011111	0337	223	0DF	0100011011	0433	283	11B
0010100100	0244	164	0A4	0011100000	0340	224	0E0	0100011100	0434	284	11C
0010100101	0245	165	0A5	0011100001	0341	225	0E1	0100011101	0435	285	11D
0010100110	0246	166	0A6	0011100010	0342	226	0E2	0100011110	0436	286	11E
0010100111	0247	167	0A7	0011100011	0343	227	0E3	0100011111	0437	287	11F
0010101000	0250	168	0A8	0011100100	0344	228	0E4	0100100000	0440	288	120
0010101001	0251	169	0A9	0011100101	0345	229	0E5	0100100001	0441	289	121
0010101010	0252	170	0AA	0011100110	0346	230	0E6	0100100010	0442	290	122
0010101011	0253	171	0AB	0011100111	0347	231	0E7	0100100011	0443	291	123
0010101100	0254	172	0AC	0011101000	0350	232	0E8	0100100100	0444	292	124
0010101101	0255	173	0AD	0011101001	0351	233	0E9	0100100101	0445	293	125
0010101110	0256	174	0AE	0011101010	0352	234	0EA	0100100110	0446	294	126
0010101111	0257	175	0AF	0011101011	0353	235	0EB	0100100111	0447	295	127
0010110000	0260	176	0B0	0011101100	0354	236	0EC	0100101000	0450	296	128
0010110001	0261	177	0B1	0011101101	0355	237	0ED	0100101001	0451	297	129
0010110010	0262	178	0B2	0011101110	0356	238	0EE	0100101010	0452	298	12A
0010110011	0263	179	0B3	0011101111	0357	239	0EF	0100101011	0453	299	12B
0010110100	0264	180	0B4	0011110000	0360	240	0F0	0100101100	0454	300	12C
0010110101	0265	181	0B5	0011110001	0361	241	0F1	0100101101	0455	301	12D
0010110110	0266	182	0B6	0011110010	0362	242	0F2	0100101110	0456	302	12E
0010110111	0267	183	0B7	0011110011	0363	243	0F3	0100101111	0457	303	12F
0010111000	0270	184	0B8	0011110100	0364	244	0F4	0100110000	0460	304	130
0010111001	0271	185	0B9	0011110101	0365	245	0F5	0100110001	0461	305	131

Binary	Octl	Decl	Hex	Binary	Octl	Decl	Hex	Binary	Octl	Decl	Hex
0100110010	0462	306	132	0101101110	0556	366	16E	0110101010	0652	426	1AA
0100110011	0463	307	133	0101101111	0557	367	16F	0110101011	0653	427	1AB
0100110100	0464	308	134	0101110000	0560	368	170	0110101100	0654	428	1AC
0100110101	0465	309	135	0101110001	0561	369	171	0110101101	0655	429	1AD
0100110110	0466	310	136	0101110010	0562	370	172	0110101110	0656	430	1AE
0100110111	0467	311	137	0101110011	0563	371	173	0110101111	0657	431	1AF
0100111000	0470	312	138	0101110100	0564	372	174	0110110000	0660	432	1B0
0100111001	0471	313	139	0101110101	0565	373	175	0110110001	0661	433	1B1
0100111010	0472	314	13A	0101110110	0566	374	176	0110110010	0662	434	1B2
0100111011	0473	315	13B	0101110111	0567	375	177	0110110011	0663	435	1B3
0100111100	0474	316	13C	0101111000	0570	376	178	0110110100	0664	436	1B4
0100111101	0475	317	13D	0101111001	0571	377	179	0110110101	0665	437	1B5
0100111110	0476	318	13E	0101111010	0572	378	17A	0110110110	0666	438	1B6
0100111111	0477	319	13F	0101111011	0573	379	17B	0110110111	0667	439	1B7
0101000000	0500	320	140	0101111100	0574	380	17C	0110111000	0670	440	1B8
0101000001	0501	321	141	0101111101	0575	381	17D	0110111001	0671	441	1B9
0101000010	0502	322	142	0101111110	0576	382	17E	0110111010	0672	442	1BA
0101000011	0503	323	143	0101111111	0577	383	17F	0110111011	0673	443	1BB
0101000100	0504	324	144	0110000000	0600	384	180	0110111100	0674	444	1BC
0101000101	0505	325	145	0110000001	0601	385	181	0110111101	0675	445	1BD
0101000110	0506	326	146	0110000010	0602	386	182	0110111110	0676	446	1BE
0101000111	0507	327	147	0110000011	0603	387	183	0110111111	0677	447	1BF
0101001000	0510	328	148	0110000100	0604	388	184	0111000000	0700	448	1C0
0101001001	0511	329	149	0110000101	0605	389	185	0111000001	0701	449	1C1
0101001010	0512	330	14A	0110000110	0606	390	186	0111000010	0702	450	1C2
0101001011	0513	331	14B	0110000111	0607	391	187	0111000011	0703	451	1C3
0101001100	0514	332	14C	0110001000	0610	392	188	0111000100	0704	452	1C4
0101001101	0515	333	14D	0110001001	0611	393	189	0111000101	0705	453	1C5
0101001110	0516	334	14E	0110001010	0612	394	18A	0111000110	0706	454	1C6
0101001111	0517	335	14F	0110001011	0613	395	18B	0111000111	0707	455	1C7
0101010000	0520	336	150	0110001100	0614	396	18C	0111001000	0710	456	1C8
0101010001	0521	337	151	0110001101	0615	397	18D	0111001001	0711	457	1C9
0101010010	0522	338	152	0110001110	0616	398	18E	0111001010	0712	458	1CA
0101010011	0523	339	153	0110001111	0617	399	18F	0111001011	0713	459	1CB
0101010100	0524	340	154	0110010000	0620	400	190	0111001100	0714	460	1CC
0101010101	0525	341	155	0110010001	0621	401	191	0111001101	0715	461	1CD
0101010110	0526	342	156	0110010010	0622	402	192	0111001110	0716	462	1CE
0101010111	0527	343	157	0110010011	0623	403	193	0111001111	0717	463	1CF
0101011000	0530	344	158	0110010100	0624	404	194	0111010000	0720	464	1D0
0101011001	0531	345	159	0110010101	0625	405	195	0111010001	0721	465	1D1
0101011010	0532	346	15A	0110010110	0626	406	196	0111010010	0722	466	1D2
0101011011	0533	347	15B	0110010111	0627	407	197	0111010011	0723	467	1D3
0101011100	0534	348	15C	0110011000	0630	408	198	0111010100	0724	468	1D4
0101011101	0535	349	15D	0110011001	0631	409	199	0111010101	0725	469	1D5
0101011110	0536	350	15E	0110011010	0632	410	19A	0111010110	0726	470	1D6
0101011111	0537	351	15F	0110011011	0633	411	19B	0111010111	0727	471	1D7
0101100000	0540	352	160	0110011100	0634	412	19C	0111011000	0730	472	1D8
0101100001	0541	353	161	0110011101	0635	413	19D	0111011001	0731	473	1D9
0101100010	0542	354	162	0110011110	0636	414	19E	0111011010	0732	474	1DA
0101100011	0543	355	163	0110011111	0637	415	19F	0111011011	0733	475	1DB
0101100100	0544	356	164	0110100000	0640	416	1A0	0111011100	0734	476	1DC
0101100101	0545	357	165	0110100001	0641	417	1A1	0111011101	0735	477	1DD
0101100110	0546	358	166	0110100010	0642	418	1A2	0111011110	0736	478	1DE
0101100111	0547	359	167	0110100011	0643	419	1A3	0111011111	0737	479	1DF
0101101000	0550	360	168	0110100100	0644	420	1A4	0111100000	0740	480	1E0
0101101001	0551	361	169	0110100101	0645	421	1A5	0111100001	0741	481	1E1
0101101010	0552	362	16A	0110100110	0646	422	1A6	0111100010	0742	482	1E2
0101101011	0553	363	16B	0110100111	0647	423	1A7	0111100011	0743	483	1E3
0101101100	0554	364	16C	0110101000	0650	424	1A8	0111100100	0744	484	1E4
0101101101	0555	365	16D	0110101001	0651	425	1A9	0111100101	0745	485	1E5

Binary	Octl	Decl	Hex	Binary	Octl	Decl	Hex	Binary	Octl	Decl	Hex
0111100110	0746	486	1E6	1000100010	1042	546	222	1001011110	1136	606	25E
0111100111	0747	487	1E7	1000100011	1043	547	223	1001011111	1137	607	25F
0111101000	0750	488	1E8	1000100100	1044	548	224	1001100000	1140	608	260
0111101001	0751	489	1E9	1000100101	1045	549	225	1001100001	1141	609	261
0111101010	0752	490	1EA	1000100110	1046	550	226	1001100010	1142	610	262
0111101011	0753	491	1EB	1000100111	1047	551	227	1001100011	1143	611	263
0111101100	0754	492	1EC	1000101000	1050	552	228	1001100100	1144	612	264
0111101101	0755	493	1ED	1000101001	1051	553	229	1001100101	1145	613	265
0111101110	0756	494	1EE	1000101010	1052	554	22A	1001100110	1146	614	266
0111101111	0757	495	1EF	1000101011	1053	555	22B	1001100111	1147	615	267
0111110000	0760	496	1F0	1000101100	1054	556	22C	1001101000	1150	616	268
0111110001	0761	497	1F1	1000101101	1055	557	22D	1001101001	1151	617	269
0111110010	0762	498	1F2	1000101110	1056	558	22E	1001101010	1152	618	26A
0111110011	0763	499	1F3	1000101111	1057	559	22F	1001101011	1153	619	26B
0111110100	0764	500	1F4	1000110000	1060	560	230	1001101100	1154	620	26C
0111110101	0765	501	1F5	1000110001	1061	561	231	1001101101	1155	621	26D
0111110110	0766	502	1F6	1000110010	1062	562	232	1001101110	1156	622	26E
0111110111	0767	503	1F7	1000110011	1063	563	233	1001101111	1157	623	26F
0111111000	0770	504	1F8	1000110100	1064	564	234	1001110000	1160	624	270
0111111001	0771	505	1F9	1000110101	1065	565	235	1001110001	1161	625	271
0111111010	0772	506	1FA	1000110110	1066	566	236	1001110010	1162	626	272
0111111011	0773	507	1FB	1000110111	1067	567	237	1001110011	1163	627	273
0111111100	0774	508	1FC	1000111000	1070	568	238	1001110100	1164	628	274
0111111101	0775	509	1FD	1000111001	1071	569	239	1001110101	1165	629	275
0111111110	0776	510	1FE	1000111010	1072	570	23A	1001110110	1166	630	276
0111111111	0777	511	1FF	1000111011	1073	571	23B	1001110111	1167	631	277
1000000000	1000	512	200	1000111100	1074	572	23C	1001111000	1170	632	278
1000000001	1001	513	201	1000111101	1075	573	23D	1001111001	1171	633	279
1000000010	1002	514	202	1000111110	1076	574	23E	1001111010	1172	634	27A
1000000011	1003	515	203	1000111111	1077	575	23F	1001111011	1173	635	27B
1000000100	1004	516	204	1001000000	1100	576	240	1001111100	1174	636	27C
1000000101	1005	517	205	1001000001	1101	577	241	1001111101	1175	637	27D
1000000110	1006	518	206	1001000010	1102	578	242	1001111110	1176	638	27E
1000000111	1007	519	207	1001000011	1103	579	243	1001111111	1177	639	27F
1000001000	1010	520	208	1001000100	1104	580	244	1010000000	1200	640	280
1000001001	1011	521	209	1001000101	1105	581	245	1010000001	1201	641	281
1000001010	1012	522	20A	1001000110	1106	582	246	1010000010	1202	642	282
1000001011	1013	523	20B	1001000111	1107	583	247	1010000011	1203	643	283
1000001100	1014	524	20C	1001001000	1110	584	248	1010000100	1204	644	284
1000001101	1015	525	20D	1001001001	1111	585	249	1010000101	1205	645	285
1000001110	1016	526	20E	1001001010	1112	586	24A	1010000110	1206	646	286
1000001111	1017	527	20F	1001001011	1113	587	24B	1010000111	1207	647	287
1000010000	1020	528	210	1001001100	1114	588	24C	1010001000	1210	648	288
1000010001	1021	529	211	1001001101	1115	589	24D	1010001001	1211	649	289
1000010010	1022	530	212	1001001110	1116	590	24E	1010001010	1212	650	28A
1000010011	1023	531	213	1001001111	1117	591	24F	1010001011	1213	651	28B
1000010100	1024	532	214	1001010000	1120	592	250	1010001100	1214	652	28C
1000010101	1025	533	215	1001010001	1121	593	251	1010001101	1215	653	28D
1000010110	1026	534	216	1001010010	1122	594	252	1010001110	1216	654	28E
1000010111	1027	535	217	1001010011	1123	595	253	1010001111	1217	655	28F
1000011000	1030	536	218	1001010100	1124	596	254	1010010000	1220	656	290
1000011001	1031	537	219	1001010101	1125	597	255	1010010001	1221	657	291
1000011010	1032	538	21A	1001010110	1126	598	256	1010010010	1222	658	292
1000011011	1033	539	21B	1001010111	1127	599	257	1010010011	1223	659	293
1000011100	1034	540	21C	1001011000	1130	600	258	1010010100	1224	660	294
1000011101	1035	541	21D	1001011001	1131	601	259	1010010101	1225	661	295
1000011110	1036	542	21E	1001011010	1132	602	25A	1010010110	1226	662	296
1000011111	1037	543	21F	1001011011	1133	603	25B	1010010111	1227	663	297
1000100000	1040	544	220	1001011100	1134	604	25C	1010011000	1230	664	298
1000100001	1041	545	221	1001011101	1135	605	25D	1010011001	1231	665	299

Binary	Octl	Decl	Hex	Binary	Octl	Decl	Hex	Binary	Octl	Decl	Hex
1010011010	1232	666	29A	1011010110	1326	726	2D6	1100010010	1422	786	312
1010011011	1233	667	29B	1011010111	1327	727	2D7	1100010011	1423	787	313
1010011100	1234	668	29C	1011011000	1330	728	2D8	1100010100	1424	788	314
1010011101	1235	669	29D	1011011001	1331	729	2D9	1100010101	1425	789	315
1010011110	1236	670	29E	1011011010	1332	730	2DA	1100010110	1426	790	316
1010011111	1237	671	29F	1011011011	1333	731	2DB	1100010111	1427	791	317
1010100000	1240	672	2A0	1011011100	1334	732	2DC	1100011000	1430	792	318
1010100001	1241	673	2A1	1011011101	1335	733	2DD	1100011001	1431	793	319
1010100010	1242	674	2A2	1011011110	1336	734	2DE	1100011010	1432	794	31A
1010100011	1243	675	2A3	1011011111	1337	735	2DF	1100011011	1433	795	31B
1010100100	1244	676	2A4	1011100000	1340	736	2E0	1100011100	1434	796	31C
1010100101	1245	677	2A5	1011100001	1341	737	2E1	1100011101	1435	797	31D
1010100110	1246	678	2A6	1011100010	1342	738	2E2	1100011110	1436	798	31E
1010100111	1247	679	2A7	1011100011	1343	739	2E3	1100011111	1437	799	31F
1010101000	1250	680	2A8	1011100100	1344	740	2E4	1100100000	1440	800	320
1010101001	1251	681	2A9	1011100101	1345	741	2E5	1100100001	1441	801	321
1010101010	1252	682	2AA	1011100110	1346	742	2E6	1100100010	1442	802	322
1010101011	1253	683	2AB	1011100111	1347	743	2E7	1100100011	1443	803	323
1010101100	1254	684	2AC	1011101000	1350	744	2E8	1100100100	1444	804	324
1010101101	1255	685	2AD	1011101001	1351	745	2E9	1100100101	1445	805	325
1010101110	1256	686	2AE	1011101010	1352	746	2EA	1100100110	1446	806	326
1010101111	1257	687	2AF	1011101011	1353	747	2EB	1100100111	1447	807	327
1010110000	1260	688	2B0	1011101100	1354	748	2BC	1100101000	1450	808	328
1010110001	1261	689	2B1	1011101101	1355	749	2ED	1100101001	1451	809	329
1010110010	1262	690	2B2	1011101110	1356	750	2EE	1100101010	1452	810	32A
1010110011	1263	691	2B3	1011101111	1357	751	2EF	1100101011	1453	811	32B
1010110100	1264	692	2B4	1011110000	1360	752	2F0	1100101100	1454	812	32C
1010110101	1265	693	2B5	1011110001	1361	753	2F1	1100101101	1455	813	32D
1010110110	1266	694	2B6	1011110010	1362	754	2F2	1100101110	1456	814	32E
1010110111	1267	695	2B7	1011110011	1363	755	2F3	1100101111	1457	815	32F
1010111000	1270	696	2B8	1011110100	1364	756·2F4		1100110000	1460	816	330
1010111001	1271	697	2B9	1011110101	1365	757	2F5	1100110001	1461	817	331
1010111010	1272	698	2BA	1011110110	1366	758	2F6	1100110010	1462	818	332
1010111011	1273	699	2BB	1011110111	1367	759	2F7	1100110011	1463	819	333
1010111100	1274	700	2BC	1011111000	1370	760	2F8	1100110100	1464	820	334
1010111101	1275	701	2BD	1011111001	1371	761	2F9	1100110101	1465	821	335
1010111110	1276	702	2BE	1011111010	1372	762	2FA	1100110110	1466	822	336
1010111111	1277	703	2BF	1011111011	1373	763	2FB	1100110111	1467	823	337
1011000000	1300	704	2C0	1011111100	1374	764	2FC	1100111000	1470	824	338
1011000001	1301	705	2C1	1011111101	1375	765	2FD	1100111001	1471	825	339
1011000010	1302	706	2C2	1011111110	1376	766	2FE	1100111010	1472	826	33A
1011000011	1303	707	2C3	1011111111	1377	767	2FF	1100111011	1473	827	33B
1011000100	1304	708	2C4	1100000000	1400	768	300	1100111100	1474	828	33C
1011000101	1305	709	2C5	1100000001	1401	769	301	1100111101	1475	829	33D
1011000110	1306	710	2C6	1100000010	1402	770	302	1100111110	1476	830	33E
1011000111	1307	711	2C7	1100000011	1403	771	303	1100111111	1477	831	33F
1011001000	1310	712	2C8	1100000100	1404	772	304	1101000000	1500	832	340
1011001001	1311	713	2C9	1100000101	1405	773	305	1101000001	1501	833	341
1011001010	1312	714	2CA	1100000110	1406	774	306	1101000010	1502	834	342
1011001011	1313	715	2CB	1100000111	1407	775	307	1101000011	1503	835	343
1011001100	1314	716	2CC	1100001000	1410	776	308	1101000100	1504	836	344
1011001101	1315	717	2CD	1100001001	1411	777	309	1101000101	1505	837	345
1011001110	1316	718	2CE	1100001010	1412	778	30A	1101000110	1506	838	346
1011001111	1317	719	2CF	1100001011	1413	779	30B	1101000111	1507	839	347
1011010000	1320	720	2D0	1100001100	1414	780	30C	1101001000	1510	840	348
1011010001	1321	721	2D1	1100001101	1415	781	30D	1101001001	1511	841	349
1011010010	1322	722	2D2	1100001110	1416	782	30E	1101001010	1512	842	34A
1011010011	1323	723	2D3	1100001111	1417	783	30F	1101001011	1513	843	34B
1011010100	1324	724	2D4	1100010000	1420	784	310	1101001100	1514	844	34C
1011010101	1325	725	2D5	1100010001	1421	785	311	1101001101	1515	845	34D

Binary	Octl	Decl	Hex	Binary	Octl	Decl	Hex	Binary	Octl	Decl	Hex
1101001110	1516	846	34E	1110001010	1612	906	38A	1111000110	1706	966	3C6
1101001111	1517	847	34F	1110001011	1613	907	38B	1111000111	1707	967	3C7
1101010000	1520	848	350	1110001100	1614	908	38C	1111001000	1710	968	3C8
1101010001	1521	849	351	1110001101	1615	909	38D	1111001001	1711	969	3C9
1101010010	1522	850	352	1110001110	1616	910	38E	1111001010	1712	970	3CA
1101010011	1523	851	353	1110001111	1617	911	38F	1111001011	1713	971	3CB
1101010100	1524	852	354	1110010000	1620	912	390	1111001100	1714	972	3CC
1101010101	1525	853	355	1110010001	1621	913	391	1111001101	1715	973	3CD
1101010110	1526	854	356	1110010010	1622	914	392	1111001110	1716	974	3CE
1101010111	1527	855	357	1110010011	1623	915	393	1111001111	1717	975	3CF
1101011000	1530	856	358	1110010100	1624	916	394	1111010000	1720	976	3D0
1101011001	1531	857	359	1110010101	1625	917	395	1111010001	1721	977	3D1
1101011010	1532	858	35A	1110010110	1626	918	396	1111010010	1722	978	3D2
1101011011	1533	859	35B	1110010111	1627	919	397	1111010011	1723	979	3D3
1101011100	1534	860	35C	1110011000	1630	920	398	1111010100	1724	980	3D4
1101011101	1535	861	35D	1110011001	1631	921	399	1111010101	1725	981	3D5
1101011110	1536	862	35E	1110011010	1632	922	39A	1111010110	1726	982	3D6
1101011111	1537	863	35F	1110011011	1633	923	39B	1111010111	1727	983	3D7
1101100000	1540	864	360	1110011100	1634	924	39C	1111011000	1730	984	3D8
1101100001	1541	865	361	1110011101	1635	925	39D	1111011001	1731	985	3D9
1101100010	1542	866	362	1110011110	1636	926	39E	1111011010	1732	986	3DA
1101100011	1543	867	363	1110011111	1637	927	39F	1111011011	1733	987	3DB
1101100100	1544	868	364	1110100000	1640	928	3A0	1111011100	1734	988	3DC
1101100101	1545	869	365	1110100001	1641	929	3A1	1111011101	1735	989	3DD
1101100110	1546	870	366	1110100010	1642	930	3A2	1111011110	1736	990	3DE
1101100111	1547	871	367	1110100011	1643	931	3A3	1111011111	1737	991	3DF
1101101000	1550	872	368	1110100100	1644	932	3A4	1111100000	1740	992	3E0
1101101001	1551	873	369	1110100101	1645	933	3A5	1111100001	1741	993	3E1
1101101010	1552	874	36A	1110100110	1646	934	3A6	1111100010	1742	994	3E2
1101101011	1553	875	36B	1110100111	1647	935	3A7	1111100011	1743	995	3E3
1101101100	1554	876	36C	1110101000	1650	936	3A8	1111100100	1744	996	3E4
1101101101	1555	877	36D	1110101001	1651	937	3A9	1111100101	1745	997	3E5
1101101110	1556	878	36E	1110101010	1652	938	3AA	1111100110	1746	998	3E6
1101101111	1557	879	36F	1110101011	1653	939	3AB	1111100111	1747	999	3E7
1101110000	1560	880	370	1110101100	1654	940	3AC	1111101000	1750	1000	3E8
1101110001	1561	881	371	1110101101	1655	941	3AD	1111101001	1751	1001	3E9
1101110010	1562	882	372	1110101110	1656	942	3AE	1111101010	1752	1002	3EA
1101110011	1563	883	373	1110101111	1657	943	3AF	1111101011	1753	1003	3EB
1101110100	1564	884	374	1110110000	1660	944	3B0	1111101100	1754	1004	3EC
1101110101	1565	885	375	1110110001	1661	945	3B1	1111101101	1755	1005	3ED
1101110110	1566	886	376	1110110010	1662	946	3B2	1111101110	1756	1006	3EE
1101110111	1567	887	377	1110110011	1663	947	3B3	1111101111	1757	1007	3EF
1101111000	1570	888	378	1110110100	1664	948	3B4	1111110000	1760	1008	3F0
1101111001	1571	889	379	1110110101	1665	949	3B5	1111110001	1761	1009	3F1
1101111010	1572	890	37A	1110110110	1666	950	3B6	1111110010	1762	1010	3F2
1101111011	1573	891	37B	1110110111	1667	951	3B7	1111110011	1763	1011	3F3
1101111100	1574	892	37C	1110111000	1670	952	3B8	1111110100	1764	1012	3F4
1101111101	1575	893	37D	1110111001	1671	953	3B9	1111110101	1765	1013	3F5
1101111110	1576	894	37E	1110111010	1672	954	3BA	1111110110	1766	1014	3F6
1101111111	1577	895	37F	1110111011	1673	955	3BB	1111110111	1767	1015	3F7
1110000000	1600	896	380	1110111100	1674	956	3BC	1111111000	1770	1016	3F8
1110000001	1601	897	381	1110111101	1675	957	3BD	1111111001	1771	1017	3F9
1110000010	1602	898	382	1110111110	1676	958	3BE	1111111010	1772	1018	3FA
1110000011	1603	899	383	1110111111	1677	959	3BF	1111111011	1773	1019	3FB
1110000100	1604	900	384	1111000000	1700	960	3C0	1111111100	1774	1020	3FC
1110000101	1605	901	385	1111000001	1701	961	3C1	1111111101	1775	1021	3FD
1110000110	1606	902	386	1111000010	1702	962	3C2	1111111110	1776	1022	3FE
1110000111	1607	903	387	1111000011	1703	963	3C3	1111111111	1777	1023	3FF
1110001000	1610	904	388	1111000100	1704	964	3C4				
1110001001	1611	905	389	1111000101	1705	965	3C5				

APPENDIX **D**

Two's Complement Number Conversion Chart

2's Comp	Decl	2's Comp	Decl	2's Comp	Decl	2's Comp	Decl
11111111	-1	11011111	-33	10111111	-65	10011111	-97
11111110	-2	11011110	-34	10111110	-66	10011110	-98
11111101	-3	11011101	-35	10111101	-67	10011101	-99
11111100	-4	11011100	-36	10111100	-68	10011100	-100
11111011	-5	11011011	-37	10111011	-69	10011011	-101
11111010	-6	11011010	-38	10111010	-70	10011010	-102
11111001	-7	11011001	-39	10111001	-71	10011001	-103
11111000	-8	11011000	-40	10111000	-72	10011000	-104
11110111	-9	11010111	-41	10110111	-73	10010111	-105
11110110	-10	11010110	-42	10110110	-74	10010110	-106
11110101	-11	11010101	-43	10110101	-75	10010101	-107
11110100	-12	11010100	-44	10110100	-76	10010100	-108
11110011	-13	11010011	-45	10110011	-77	10010011	-109
11110010	-14	11010010	-46	10110010	-78	10010010	-110
11110001	-15	11010001	-47	10110001	-79	10010001	-111
11110000	-16	11010000	-48	10110000	-80	10010000	-112
11101111	-17	11001111	-49	10101111	-81	10001111	-113
11101110	-18	11001110	-50	10101110	-82	10001110	-114
11101101	-19	11001101	-51	10101101	-83	10001101	-115
11101100	-20	11001100	-52	10101100	-84	10001100	-116
11101011	-21	11001011	-53	10101011	-85	10001011	-117
11101010	-22	11001010	-54	10101010	-86	10001010	-118
11101001	-23	11001001	-55	10101001	-87	10001001	-119
11101000	-24	11001000	-56	10101000	-88	10001000	-120
11100111	-25	11000111	-57	10100111	-89	10000111	-121
11100110	-26	11000110	-58	10100110	-90	10000110	-122
11100101	-27	11000101	-59	10100101	-91	10000101	-123
11100100	-28	11000100	-60	10100100	-92	10000100	-124
11100011	-29	11000011	-61	10100011	-93	10000011	-125
11100010	-30	11000010	-62	10100010	-94	10000010	-126
11100001	-31	11000001	-63	10100001	-95	10000001	-127
11100000	-32	11000000	-64	10100000	-96	10000000	-128

Index